3rd International Conference on Manufacturing Technologies (ICMT 2019)

IOP Conference Series: Materials Science and Engineering
Volume 521

San Francisco, California, USA
4 – 7 January 2019

ISBN: 978-1-5108-8739-8
ISSN: 1757-8981

Printed from e-media with permission by:

Curran Associates, Inc.
57 Morehouse Lane
Red Hook, NY 12571

CURRAN ASSOCIATES INC. proceedings.com

Some format issues inherent in the e-media version may also appear in this print version.

Printed by Curran Associates, Inc. (2019)

For permission requests, please contact the Institute of Physics at the address below.

Institute of Physics
Dirac House, Temple Back
Bristol BS1 6BE UK

Phone: 44 1 17 929 7481
Fax: 44 1 17 920 0979

techtracking@iop.org

Additional copies of this publication are available from:

Curran Associates, Inc.
57 Morehouse Lane
Red Hook, NY 12571 USA
Phone: 845-758-0400
Fax: 845-758-2633
Email: curran@proceedings.com
Web: www.proceedings.com

Table of contents

Volume 521

2019 3rd International Conference on Manufacturing Technologies (ICMT)

4–7 January 2019, San Francisco, CA, USA

Accepted papers received: 22 March 2019
Published online: Pl Online DATE

Preface

Papers

IOP Conf. Series: Materials Science and Engineering **521** (2019) 011001 doi:10.1088/1757-899X/521/1/011001

Preface

The 2019 3rd International Conference on Manufacturing Technologies (ICMT2019) was held in held in San Francisco, USA during January 4-7, 2019. This volume consists of selected papers presented at ICMT 2019, which are addressing diverse topics related to recent trends in the inter-related areas of Mechanical, Manufacturing and Materials Engineering. The conference provides opportunities for the delegates to exchange new ideas and application through face-to-face discussions, to establish business or research relations and to find global partners for future collaborations. We hope that you will find its contents interesting, stimulating and educative. ICMT 2019 is organized by the South Asia Institute of Science and Engineering.

The technical objective of the conference is to address key issues associated with science and technology in these rapidly evolving fields of research and to promote contact between basic researcher and technological needs for real industrial applications.

The conference program offers invited, oral and poster presentations. The invited talks are focused on the potential application of mechanical, manufacturing, mechatronics and materials engineering in various fields. Many excellent manuscripts were selected after peer review and were recommended for publication in this volume. Topics cover all aspects of the areas listed above.

I would like to thank the members of the conference committee, the reviewers who spared their valuable time, for their advice which have certainly helped to improve the quality, accuracy, relevance and to enhance the technical contributions of each paper selected for this conference program and volume for publication. And thank you to each author and participant who has contributed to the success of the conference.

Prof. Dr. Ridha Ben Mrad
University of Toronto, Canada
January/4-7/2019

IOP Conf. Series: Materials Science and Engineering **521** (2019) 011002 doi:10.1088/1757-899X/521/1/011002

CONFERENCE COMMITTEES

Conference Chairs

Prof. Ramesh K. Agarwal, Washington University in St. Louis, USA
Prof. Dr. Ridha Ben Mrad, University of Toronto, Canada

Program Committee Chairs

Prof. Devki N. Talwar, Indiana University of Pennsylvania, USA
Prof. Yong Suk Yang, Pusan National University, Korea
Prof. TANG Guoyi, Advanced Materials Institute of Graduate School at Shenzhen, Tsinghua University, China

International Technical Committees

Prof. K, Lawrence DeVries,The University of Utah, USA
Prof. Guo Xinli, Southeast University, China
Prof. Qing-Qing Ni, Shinshu University, Japan
Prof. Md Mustafizur Rahman, Universiti Malaysia Pahang, Malaysia
Prof. Sima Habibi, Islamic Azad University Of Yadegare Imam Khomeini Rah, Iran
Prof. Faiz Mohammad, Aligarh Muslim University, India
Prof. Dr. L. P. Purohit, Gurukula Kangri University, India
Assoc. Prof. Jin-Soo Park, Sangmyung University, South Korea
Assoc. Prof. Recai KUS, Selcuk University, Turkey
Assoc. Prof. Anselmo Alves Bandeira, Polytechnic School of the Federal University of Bahia - UFBA, Brazil
Dr. Xiaoyi Xu, Stanford university, USA
Dr. Ke Chang, Applied Materials Pte. Ltd.
Dr. Dongsoo Kim, Korea Institute of Materials Science, Korea
Wenke Gao, Lanzhou University of Technology, China
Anselmo Alves Bandeira, Federal University of Bahia, Brazil
M.P Jahan, Miami University, USA

IOP Conf. Series: Materials Science and Engineering **521** (2019) 011003 doi:10.1088/1757-899X/521/1/011003

Peer review statement

All papers published in this volume of *IOP Conference Series: Materials Science and Engineering* have been peer reviewed through processes administered by the proceedings Editors. Reviews were conducted by expert referees to the professional and scientific standards expected of a proceedings journal published by IOP Publishing.

IOP Conf. Series: Materials Science and Engineering **521** (2019) 012001 doi:10.1088/1757-899X/521/1/012001

Effect of Laser Beam Conditioning on Fabrication of Micro-Channels in Al_2O_3 Bio-ceramics Using Nd:YAG Laser

Basem M. A. Abdo[1,2a], Abdulaziz Al-Tamimi[1], Abdulrahman Al-Ahmari[1,2], Hisham Alkhalefah[2]

[1]Industrial Engineering Department, College of Engineering, King Saud University, Riyadh, Saudi Arabia
[2]Princess Fatima Alnijris's Research Chair for Advanced Manufacturing Technology (FARCAMT), Advanced Manufacturing Institute, King Saud University, Riyadh, Saudi Arabia

Corresponding author: babdo@ksu.edu.sa

Abstract. In the present study, laser micro-milling tests were carried out to fabricate micro-channels on Alumina bio-ceramics (Al_2O_3), using a Q-Switched 30W Nd:YAG pulsed laser. A systematic approach based on a full factorial Design of Experiment (DoE) has been successfully applied with the aim to detect which and how the key input laser process parameters affect the channel dimensional accuracy. The examined process parameters were the laser beam scanning speed, the pulse frequency and the pulse intensity. Optical microscope was used to analyze the channel geometries responses (i.e. channel's top width, bottom width, depth, and taper wall angle). Moreover, mathematical models for predicting the micro-channel geometries are successfully proposed for controlled micro-milling of micro-channels in Al_2O_3. Results reveal that, the change of scanning speed and laser intensity significantly affected the ablated channel's geometries. Further it is observed that the channel depth and width increase linearly with increasing of laser intensity and decreasing of scanning speed and not much affected by changing of pulse frequency. Finally, the experimental results bear a good agreement with the proposed prediction models.

1. Introduction

Bio-ceramics such as alumina, zirconia etc. have been used in many applications in medical field such as micro-fluidics, lab-on-chip, dental restoration, etc. that rely on precision machined micro-features (e.g. micro channels, micro holes). This is duo to their superior properties such as high hardness, high strength at elevated temperatures, low density and resistance to wear, creep and chemical degradation. However, duo to their excellent properties, bio-ceramics considered as difficult to machine materials [1]. High hardness and low ductility properties make bio-materials difficult to fabricate by traditional processes such as milling and grinding. Moreover, properties of resistance to chemical degradation, and poor electrical conductivity acts as impediment in machining of bio-materials via chemical machining and/or electric discharge machining processes [2,3].

As a result, there exist considerable impediments to market expansion of such materials. One of the primary impediments is high cost of machining with current technologies [4,5].

Laser micro-milling process (LMMP) provides a new method of producing micro-parts in a wide range of materials directly from CAD data. It has rapidly become inexpensive that allows the production of parts with complex shapes without expensive tooling, and it offers effective and controllable machining for most advanced applications. Compared with other conventional processes (milling, drilling), LMMP has several advantages such as absence of tool related issues, capability to machine all types of materials irrespective of their hardness, thermal and electrical conductivity and capability of easy and fast micro machining [6,7]. In LMMP, the material is removed layer by layer due to the ablation operation causing vaporization of material as a result of interaction between a high energy laser beam and the work piece being machined.

Micro-machining of bio-ceramics has been attempted by many researchers using various types of laser processes. The main focus is to achieve the desired geometries of micro-features such as micro-channels, micro-holes, micro-grooves etc. For example, Samant. & Dahotre [1] reviewed the literatures about laser micromachining of structure ceramics, and pointed out that the mechanism governing the material removal is a function of the material properties and the laser processing conditions. Fabrication of micro-channels and micro-pits on ceramic (Al_2O_3) and steel (100Cr6) surfaces is reported in [8]. Wherein Nd:YAG-laser is used to reduce the coefficient of friction of the machined substrates. In the study documented in [9], femtosecond pulsed laser was employed to drill blind and through holes in titanium carbide ceramic. The results revealed that laser fluence and repetition rate had more impact on ablation efficiency and quality of the machined holes. Developing a model to investigate the parametric effects of femtosecond laser ablation on machining of Al_2O_3 ceramics was also found in [10] .The authors in [11] used nanosecond laser to fabricate precise micro feature on yttria-tetragonal zirconia polycrystal ceramic. Different laser parameters have been investigated in this study in order to achieve high quality micro-sized step (500μm width 100μm depth) and blind hole (500μm dia). Results revealed that 1.35 mm^3/min material removal rate and the smooth machined surface with Ra =2.824μm roughness can be achieved. Fabrication of micro-scale textured grooves on the green ZrO_2 ceramic surfaces by Nd:YAG laser ablation is reported in [12] where the authors evaluated the effects of laser speed, frequency, and power on surface quality and groove dimensions. Results revealed that, better surface quality – of the grooves with a width of 30–50 μm and a depth of 15–50 μm - can be obtained with frequency below 40 Hz, power below 6 W, and scanning velocity above 200 mm/s. M.M. Hanon et al. [13] investigated the effects of laser parameters such as peak power, pulse duration, focal position and repetition rate on the dimensions and microstructure of the micro-holes drilled on alumina ceramic using Nd:YAG laser. Results indicated that, for a single laser pulse application - peak power and pulse duration can be effectively used for controlling the crater depth without formation of any defects. Moreover, crater depth increased with the number of pulses due to insufficient recoil pressure inside the cavity.

From the above literature survey, it can be seen that very few studies have been reported on machining of the micro-channels in bio-ceramics. Moreover, no research found in open literature considered the effects of laser parameters on all channel dimensional characteristics. In the present study, laser micro-milling tests were carried out to fabricate micro-channels on alumina ceramic materials by using a Q-Switched 30 W Nd:YAG laser. A full factorial Design of Experiment (DoE) has been successfully applied with the aim to detect which and how the key laser input process parameters affect the channel dimensional accuracy. The examined input parameters were the laser beam scanning speed, the pulse frequency and the laser intensity. The responses were selected based on channels geometries includes channel depth (D), top width (TW), bottom width (BW) and wall taper angle (θ). Single factor effects (main effects) as well as interaction effects of laser parameters have been analyzed. Finally, mathematical model based on DoE regression was successfully applied to predict all responses.

2. Materials and methods

Alumina ceramic (Al_2O_3) was used as the workpiece material from CeramTec Germany. Each of the samples were of the dimension 50 mm × 50 mm ×10 mm approximately. The mechanical and thermal properties is presented in Table 1.

IOP Conf. Series: Materials Science and Engineering **521** (2019) 012001 doi:10.1088/1757-899X/521/1/012001

Table 1: Mechanical and thermal properties of as received alumina materials

Property	Unit	Value
Flexural strength	MPa	345
Vickers hardness (HV 0,5)	..	2000
Young's modulus	GPa	378
Density	g/cm^3	≥ 3.97
Thermal expansion coefficient WAK 20-1000°C	$10^{-6}\ K^{-1}$	6.7
Specific heat cp 20°C	kJ/kgK	0.9
Thermal conductivity 25°C	W/mK	20-30

Lasertec-40 from DMG Mori Sieki (see Fig. 1) is used for micro-milling of alumina samples. It is continuous wave pulsed laser with 1064 nm wavelength capable of generating train of pulses with 10 µm pulse duration following the Gaussian mode (TEM00) with average and power of 30 W. The laser unit mainly consists of an Nd:YAG (neodymium-doped yttrium aluminum garnet; Nd: Y3Al5O12) rod which acts as the lasing medium, Flash lamp as a pumping source and a Q-switch for pulsing.

Figure 1. Experimental setup of Nd:YAG laser micro-milling for fabrication micro-channels on Al_2O_3

The micro-channels with square cross-sectional dimensions (100 µm × 100 µm) were used in this study. The channel dimensions refer to a commonly used micro-channel dimensions in micro-fluidics application [14–16]. Fig. 2 shows the channel geometrical dimensions analyzed to evaluate the channel accuracy in this study.

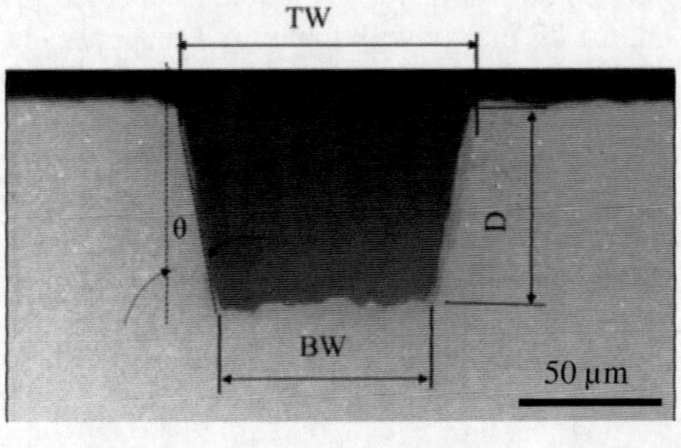

Figure 2. Channel geometrical dimensions used as responses in this study.

Three input parameters along with three levels each resulted in 27 runs carried out as a full factorial design in the current research. The laser input parameters were varied as follows, the scanning speed (Ss) in the range of 100-300 mm/s, the pulse frequency (Pf) in the range of 5- 15 kHz, and the laser intensity in the range of 88-96 %. The laser parameters and their levels were selected based on previous studies [12,17–19] and pilot runs.

IOP Conf. Series: Materials Science and Engineering **521** (2019) 012001 doi:10.1088/1757-899X/521/1/012001

Table 2 summarized the laser factors and their related levels adopted in this study. Design of experiments (DoE) based on full factorial method was carried out by using Minitab software version 17.

Table 2: Range of the laser input parameters used in the preliminary experiments

Factors	levels		
	1	2	3
Scanning speed (Ss), mm/s	100	200	300
Pulse frequency (Pf), kHz	5	10	15
Pulse intensity (Pi), %	88	92	96

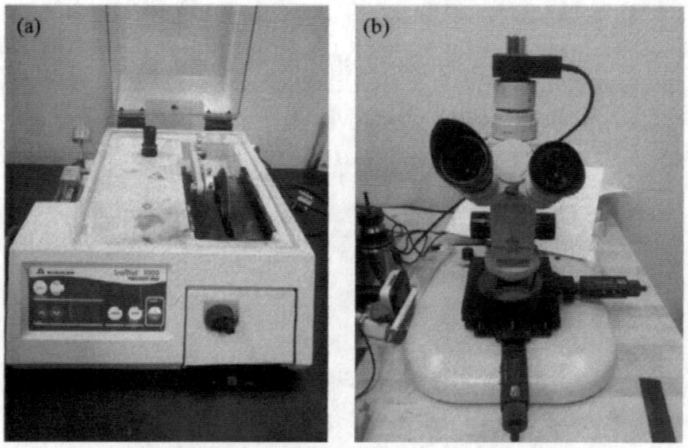

Figure 3. (a) Precision cutter used to cut the channels. (b) Optical microscope used to measure the channel dimensions

For accurate cross-sectional observations, the channels were sliced by using precision cutter (IsoMet 1000 Precision Cutter - from Buehler Co) as shown in Fig. 3a. The measurement of channels geometries was carried out by using optical microscope (mikroskop technik rathenow from ASKANIA Co.) as shown in Fig. 3b

3. Results and analysis

Design of Experiment (DOE) based on full factorial method is selected for the present study. With three factors having three level each, the 27 matrix with the output responses is listed in Table 3.

Table 3: DOE runs for the experimental plan.

Ex. #	Input parameters			Responses			
	Ss (mm/s)	Pf (kHz)	Pi (%)	D (μm)	TW (μm)	BW (μm)	θ (°)
1	300	15	92	65	93	40	12
2	100	10	88	77	95	50	17
3	100	10	92	85	86	45	16
4	200	5	96	91	98	37	18
5	300	10	96	76	92	28	23
6	300	10	92	68	90	29	12
7	300	5	96	76	95	37	20
8	100	15	88	79	90	76	11
9	200	10	96	82	95	24	15
10	200	15	92	79	93	48	21
11	200	15	96	86	92	35	19
12	100	5	88	76	95	35	20
13	300	5	92	79	86	38	23
14	300	15	88	65	97	24	11

15	300	10	88	81	84	18	23
16	100	15	96	73	96	58	17
17	200	5	88	70	92	21	10
18	200	10	92	75	94	46	12
19	100	10	96	90	102	13	17
20	200	5	92	85	94	13	13
21	300	5	88	74	84	32	11
22	200	15	88	72	86	47	17
23	100	15	92	82	96	14	16
24	100	5	92	108	99	71	21
25	300	15	96	84	91	46	18
26	100	5	96	119	94	36	19
27	200	10	88	73	89	72	15

The main and interaction effects of selected laser process parameters (namely: scanning speed (Ss), pulse frequency (Pf) and pulse intensity (Pi)) on the channels' dimensions (namely: depth (D), top width (TW), bottom width (BW) and taper angle (θ)) fabricated on alumina ceramic materials are shown in Fig. 4 and Fig. 5 respectively. As can be seen in Fig. 4(a-d), the ablated D, TW and BW decreases with increasing of laser scanning speed. This is due to, when the material is irradiated at low scan speed, the energy required to melts the material and heats it to the vaporization temperature decrease. It results in an increase of laser efficiency, and, then, in an increase of obtained depth and widths. On the other hand, the Ss has less effects on the average taper angle (θ) as shown in Fig. 4d. The pulse frequency has significant effects on the channel depth, bottom width and taper angle. However, the top width remains in fluctuating mode with increasing of pulse frequency. The channel depth (D) exists in decrease proportional with increasing of pulse frequency. This is because when the pulse frequency increases the pulse overlap decrease and this lead to decrease laser energy. Regarding to the effects of laser intensity on channel geometries, as expected, the laser intensity affects the channel depth, top width and taper angle that is directly proportional. In contrast, pulse intensity affects the bottom width in inversely proportional. This is because, laser loses its energy when it goes deeper, that why less bottom width and more taper angle can be obtained with high pulse intensity.

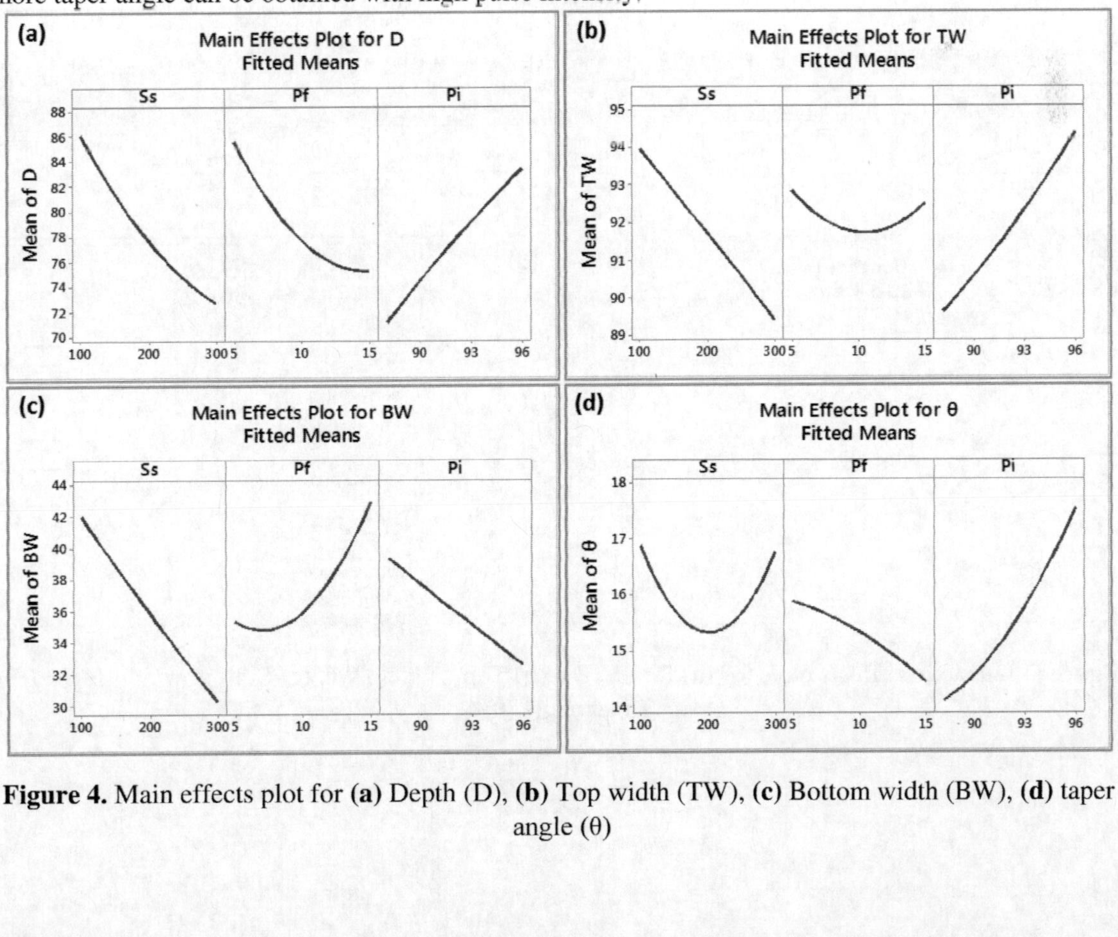

Figure 4. Main effects plot for **(a)** Depth (D), **(b)** Top width (TW), **(c)** Bottom width (BW), **(d)** taper angle (θ)

Fig. 5(a-d) shows the parametric interaction effects on the channel geometries. In general, it's found that, the interaction between scanning speed × pulse frequency (Ss × Pf) and pulse frequency × pulse intensity (Pf × Pi) have significant influence on all channel geometries (D, TW, BW and θ). Moreover, the interaction between scanning speed × pulse intensity (Ss × Pi) has more influence only on BW. Based on the DoE results, full quadratic mathematical models (Equations (1) – (4)) were applied to predict all responses (D, TW, BW, and θ). The models have been tested and verified within the input parameters range and out of the input parameters range. The verification results (see Table 4) showed that there is a sound agreement between the predicted and actual outputs within the input range. The maximum error between the predicted (out of the range) and the actual output was less than 9 %.

$$\mathbf{D} = -518 + 0.426\,\mathrm{Ss} + 9.9\,\mathrm{Pf} + 9.7\,\mathrm{Pi} + 0.000172\,\mathrm{Ss} \times \mathrm{Ss} + 0.109\,\mathrm{Pf} \times \mathrm{Pf} - 0.028\,\mathrm{Pi} \times \mathrm{Pi} + 0.00900\,\mathrm{Ss} \times \mathrm{Pf} - 0.00708\,\mathrm{Ss} \times \mathrm{Pi} - 0.162\,\mathrm{Pf} \times \mathrm{Pi} \quad (1)$$

$$\mathbf{TW} = 165 - 0.076\,\mathrm{Ss} + 2.31\,\mathrm{Pf} - 2.2\,\mathrm{Pi} - 0.000006\,\mathrm{Ss} \times \mathrm{Ss} + 0.0378\,\mathrm{Pf} \times \mathrm{Pf} + 0.017\,\mathrm{Pi} \times \mathrm{Pi} + 0.00367\,\mathrm{Ss} \times \mathrm{Pf} + 0.00021\,\mathrm{Ss} \times \mathrm{Pi} - 0.0417\,\mathrm{Pf} \times \mathrm{Pir}^2 \quad (2)$$

$$\mathbf{BW} = 394 - 1.81\,\mathrm{Ss} + 9.8\,\mathrm{Pf} - 4.0\,\mathrm{Pi} + 0.000022\,\mathrm{Ss} \times \mathrm{Ss} + 0.129\,\mathrm{Pf} \times \mathrm{Pf} + 0.003\,\mathrm{Pi} \times \mathrm{Pi} - 0.0005\,\mathrm{Ss} \times \mathrm{Pf} + 0.0190\,\mathrm{Ss} \times \mathrm{Pi} - 0.125\,\mathrm{Pf} \times \mathrm{Pi} \quad (3)$$

$$\boldsymbol{\theta} = 286 - 0.276\,\mathrm{Ss} + 0.27\,\mathrm{Pf} - 5.7\,\mathrm{Pi} + 0.000150\,\mathrm{Ss} \times \mathrm{Ss} - 0.0067\,\mathrm{Pf} \times \mathrm{Pf} + 0.031\,\mathrm{Pi} \times \mathrm{Pi} + 0.00050\,\mathrm{Ss} \times \mathrm{Pf} + 0.00229\,\mathrm{Ss} \times \mathrm{Pi} - 0.0042\,\mathrm{Pf} \times \mathrm{Pi} \quad (4)$$

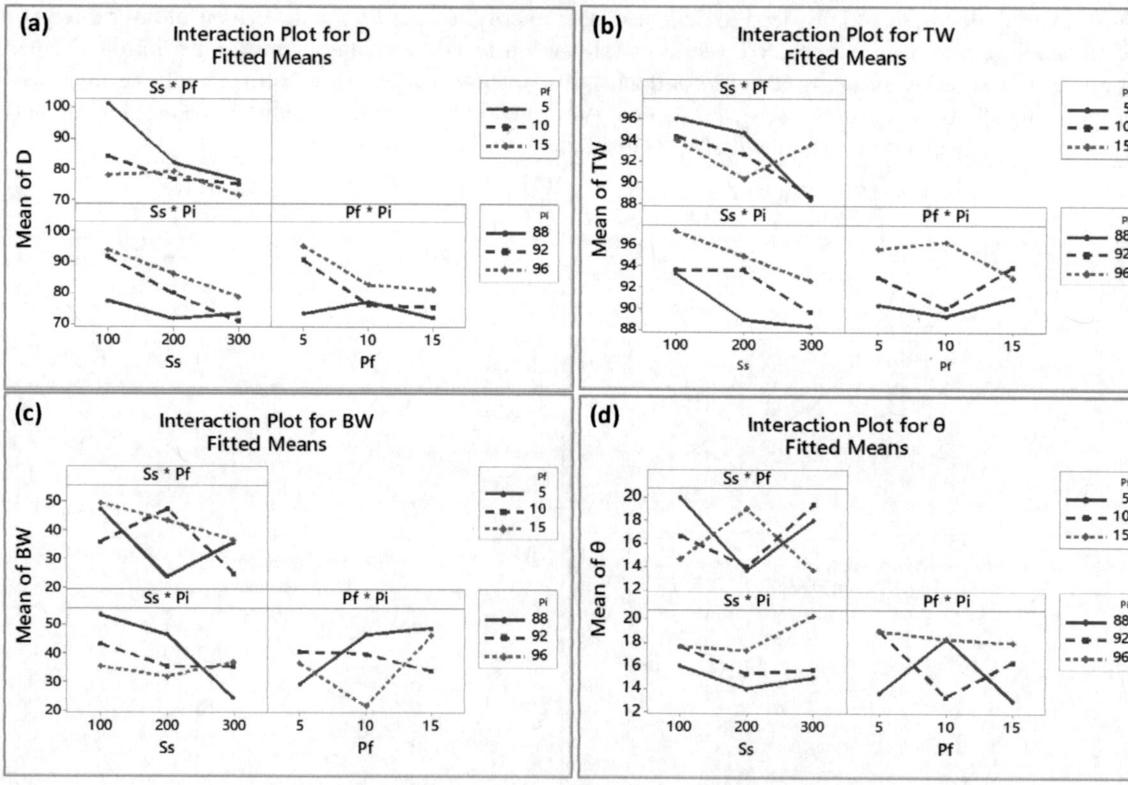

Figure 5. Interaction effects plots for **(a)** Depth (D), **(b)** Top width (TW), **(c)** Bottom width (BW), **(d)** Taper angle (θ)

Table 4. Actual experimental Vs model predicted for micro-channel geometries (verification of the models)

Models' verification range	Input parameters			Outputs							
				Actual				Predicted			
	Ss (mm/s)	Pf (kHz)	Pi (%)	D (μm)	TW (μm)	BW (μm)	θ (°)	D (μm)	TW (μm)	BW (μm)	θ (°)
Within the input range	250	12	90	97	99	87	6	98.5	101.2	82.6	7.4
	150	7	94	104	96	68	8	102.5	96.5	72.4	6.9
Out of the input range	75	18	97	107	97	64	13	111.2	97.8	67.3	12.4
	350	10	87	69	92	72	14	64.9	94.1	67.8	12.7

4. Conclusion

The Nd:YAG pulsed laser micro-milling has the capability of producing successful quality micro-channel generation on Al_2O_3.

Full factorial Design of Experiment (DoE) method has been successfully applied to detect the influence of Nd:YAG laser process parameters on the accuracy of channel geometries. Based on the experimental investigations during micro-channeling of Al_2O_3 by LMMP and the developed mathematical modes the following outcomes may possibly be drawn as listed below:

- The channel ablated depth (D) seems to be directly proportional to the laser intensity and inversely to the scanning speed.
- The channel top width (TW) was found to be closed to the design value of 100 μm at almost all levels of the input parameters.
- None of the channel bottom widths (BW) were achieved close to the designed widths. that is owing to the inherent characteristics of the laser that it imparts some taper as the machining depth progresses. The laser beam energy decreases by the time it reaches the channel bottom.
- The channel taper angle (θ) is highly effected by the pulse intensity. The channel taper angle increases with increasing of pulse intensity.
- With moderate level of pulse intensity (92%) and with appropriate combinations of other input parameters, desired channel shapes can be fabricated on Al_2O_3 forte material.
- Mathematical models for predicting the micro-channel geometries are successfully proposed for controlled micro-milling of micro-channels in Al_2O_3. The experimental results bear a good agreement with the proposed models. These models can help directly to predict the micro-channel dimensions under different parametric settings of Nd:YAG laser machining of Al_2O_3.

5. References

[1] A.N. Samant and N.B. Dahotre, Laser machining of structural ceramics — A review, *J. Eur. Ceram. Soc.* 29 (2009) 969–993.

[2] Y. Zhang et al., Edge chipping and flexural resistance of monolithic ceramics, Dent. Mater. 29 (12) (2013) 1201–1208.

[3] Z.C. Li et al., Edge-chipping reduction in rotary ultrasonic machining of ceramics: Finite element analysis and experimental verification, Int. J. Mach. Tools Manuf. 46 (12–13) (2006) 1469–1477.

[4] S. Jahanmir et al., Ceramic Machining : Assessment of Current Practice and Research Needs in the United States, NIST Spec. Publ. Gaithersburg, MD (1992) p.iii–iv.

[5] C. Barnes, P. Shrotriya, and P. Molian, Water-assisted laser thermal shock machining of alumina, Int. J. Mach. Tools Manuf. 47 (12–13) (2007) 1864–1874.

[6] S. Mishra and V. Yadava, Laser Beam MicroMachining (LBMM) – A review, Opt. Lasers Eng. 73 (2015) 89–122.

[7] N. Ahmed, S. Darwish, and A.M. Alahmari, Laser Ablation and Laser-Hybrid Ablation Processes: A Review, Mater. Manuf. Process. 31 (9) (2016) 1121–1142.

[8] S. Schreck and K. Zum Gahr, Laser-assisted structuring of ceramic and steel surfaces for improving tribological properties, Appl. Surf. Sci. 247 (2005) 616–622.

[9] Y. Zhang et al., Micromachining features of TiC ceramic by femtosecond pulsed laser, Ceram. Int. 41 (5) (2015) 6525–6533.
[10] B. Chen et al., Parametric effects on femtosecond laser ablation of Al2O3 ceramics, Ceram. Int. 39 (2013) S341–S344.
[11] J. Li et al., Precise micromachining of yttria-tetragonal zirconia polycrystal ceramic using 532 nm nanosecond laser, Ceram. Int. 42 (3) (2016) 4377–4385.
[12] Y. Liu et al., Fabrication of micro-scale textured grooves on green ZrO 2 ceramics by pulsed laser ablation, Ceram. Int. 43 (8) (2017) 6519–6531.
[13] M.M. Hanon et al., Experimental and theoretical investigation of the drilling of alumina ceramic using Nd : YAG pulsed laser, Opt. Laser Technol. 44 (4) (2012) 913–922.
[14] G.M. Whitesides, The origins and the future of microfluidics, Nature 442 (7101) (2006) 368–373.
[15] M. Farhan Shafique et al., Fabrication of embedded microfluidic channels in low temperature co-fired ceramic technology using laser machining and progressive lamination, J. Eur. Ceram. Soc. 31 (13) (2011) 2199–2204.
[16] S.E. Ong et al., Fundamental principles and applications of microfluidic systems, Front. Biosci. 13 (1) (2008) 2757–2773.
[17] S. Guarino et al., Laser milling of yttria-stabilized zirconia by using a Q-switched Yb : YAG fiber laser : experimental analysis, (2017).
[18] B.M.A. Abdo et al., Experimental investigation and multi-objective optimization of Nd:YAG laser micro-channeling process of zirconia dental ceramic, Int. J. Adv. Manuf. Technol. (2018).
[19] B.M.A. Abdo et al., Laser micro-milling of bio-lox forte ceramic: An experimental analysis, Precis. *Eng*. 53 (April) (2018) 179–193.

IOP Conf. Series: Materials Science and Engineering **521** (2019) 012002 doi:10.1088/1757-899X/521/1/012002

Stamping of a Cross-shaped Part with 5052, 5754 and 6016 Aluminum Alloy Sheets – Experimental and Finite Element Analysis Comparison

R Esmaeilpour[1], S A Nazari Tiji[1], H Kim[1], T Park[1], H Kim[1], F Pourboghrat[1,2], B Mohammed[1,3]

[1] Department of Integrated Systems Engineering, The Ohio State University, 210 Baker Systems, 1971 Neil Avenue, Columbus, OH 43210, United States
[2] Department of Mechanical and Aerospace Engineering, The Ohio State University, 201 W 19th Ave, Columbus, OH 43210, United States
[3] Department of Mechanical Engineering, Michigan State University, 428 S Shaw Ln, East Lansing, MI 48824, United States

Corresponding author e-mail: esmaeilpour.1@osu.edu

Abstract. The application of aluminum alloys is continuing to increase in automotive and aerospace industries. In this study, the stamping of a cross-shaped part, representing typical deep drawn automotive parts, was attempted with three different types of aluminum alloy sheets. To characterize the anisotropic properties of the material, uniaxial tensile tests were performed with ASTM standard dog-bone specimens cut from AA5052, AA5754 and AA6016 T4 alloy sheets. The stamping of the cross-shaped parts was conducted using a high capacity servo press with a servo-cushion system. Deformation strains developing in stamped parts were measured with the ARGUS system for comparison with finite element simulation results. Formed parts were also cut in two different directions using a water-jet machine, and thickness variations along the length of the cut section were measured. The measured surface strains and thickness distributions were compared with finite element simulations results obtained with LS-Dyna explicit code using the Yld2000-2D and Hill's 48 yield functions. Overall, experimental and numerical results correlated well, especially with the Yld2000-2D yield function, implying that this yield function is well suited for the modeling of the complex cross-shaped parts with aluminum alloys.

1. Introduction

Advanced manufacturing technology and weight reduction are two important topics in material processing when it comes to the reduction of energy consumption in transportation industry. Weight reduction can be achieved by optimizing the structural design and usage of lightweight materials, such as magnesium, titanium and aluminum alloys. The application of aluminum alloy, in particular, is increasingly applied in both automotive and aerospace industries. The average usage of aluminum in passenger cars has doubled during the last decade [1-3]. AA6016 is mostly used for the autobody application in Europe. This alloy shows a superior formability and corrosion resistance [1]. Forming behavior of AA6016 was studied by Klos et al. using Hill's 90 yield criterion. They compared cup drawing tests with numerical simulations in

IOP Conf. Series: Materials Science and Engineering **521** (2019) 012002 doi:10.1088/1757-899X/521/1/012002

order to validate the simulation of complex parts [4]. For fabricating automotive parts like fuel tank, floor panel, side and inner panels, aluminum alloy AA5754 and AA5052 are mostly used [5].
Fracture during press forming is a big issue for aluminum alloy sheets because of low ductility. Therefore, optimization of forming conditions to prevent the fracture of aluminum alloy sheets has been widely studied using finite element analysis (FEA) [6]. The FEA result accuracy depends on the constitutive model describing the behavior of the material [7]. Many constitutive models have been used to describe the material anisotropy starting with Hill's 48. During the last three decades, many anisotropic models have been used to improve the modeling of material behavior [8-15]. One of the most famous material models was introduced by Barlat as Yld2000-2D which is an anisotropic yield function for the plane stress condition [16].
In this study, tensile tests for dog bone specimens of three different aluminum alloys, 5052, 5754 and 6016 in 0, 45, and 90 degrees with respect to the rolling direction have been performed. The anisotropic behaviors of these aluminum alloy sheets have been modeled using two different yield functions, Hill's 48 and Yld2000-2D. Then, cross die forming tests were conducted using a high capacity servo press with a servo-cushion system. Strain distribution measurement was performed using ARGUS system based on conventional grid analysis. Cross formed samples were cut using a water-jet machine for thickness measurements in two directions. These experimental results were compared with simulations results using Hill's 48 and Yld2000-2D yield functions in LS-Dyna explicit code.

2. Material characterization

2.1. Uniaxial tensile tests

Uniaxial tests with standard ASTM rectangular dog-bone shaped samples of the three aluminum alloys (AA5052 and AA5754 without any heat treatment and AA6016 with T-4 temper) were performed on a MTS testing machine to obtain the stress-strain curves and anisotropic properties at room temperature. Two extensometers were used to measure the longitudinal and width strain during the test. Tensile coupons were fabricated in three different directions i.e. rolling direction (RD), diagonal direction (DD), and transverse direction (TD) to obtain the anisotropy coefficients (R_0, R_{45}, and R_{90}). The tests were repeated three times for each condition. All the tests were done at a strain rate of $0.006s^{-1}$.

2.2 Hill's 48 and Yld2000-2D anisotropy coefficients

By using the measured R_0, R_{45}, and R_{90}, the anisotropy coefficients for the Hill'48 yield function were calibrated as presented in Table 2. It should be noted that for the calibration of Hill's 1948 yield function, the material parameters for out of plane stress components were assumed to be same as the isotropic von Mises yield function, i.e. L = M = 1.5 [17,18]. In addition, the anisotropy coefficients of Yld2000-2D were calibrated using the normalized yield stress values (i.e. $\sigma_0/\sigma_0 = 1$, σ_{45}/σ_0, and σ_{90}/σ_0), as well as the plastic anisotropy values (i.e. R_0, R_{45}, and R_{90}) as listed in Table3. Note that the yield stress and R-value for the balanced biaxial stress mode were assumed to be 1.0. The measured (or assumed) data used for calibration of the anisotropic coefficients are summarized in Table 1.

Table 1. Normalized measured yield stress and R-values for AA5052, AA5754, and AA6016

Material	σ_0	σ_{45}	σ_{90}	σ_b	R_0	R_{45}	R_{90}	R_b
AA5052	1	0.951	0.931	1	0.896	0.696	0.752	1
AA5754	1	0.965	1.021	1	0.928	0.996	1.094	1
AA6016	1	0.973	0.969	1	1.240	0.720	1.150	1

Table 2. Hill's48 anisotropy coefficients for AA5052, AA5754, and AA6016

Material	F	G	H	L	M	N
AA5052	0.6282	0.5275	0.4725	1.5	1.5	1.3816
AA5754	0.4401	0.5185	0.4815	1.5	1.5	1.4345
AA6016	0.4814	0.4464	0.5536	1.5	1.5	1.1319

Table 3. Yld2000-2D anisotropy coefficients for AA5052, AA5754, and AA6016

Material	α_1	α_2	α_3	α_4	α_5	α_6	α_7	α_8
AA5052	0.934	1.092	0.870	1.044	1.012	1.013	1.009	1.169
AA5754	1.001	0.981	1.041	0.988	1.000	0.983	1.030	1.084
AA6016	0.980	1.072	0.967	1.008	0.997	1.022	0.995	1.109

3. Cross die forming process

The cross die forming test was conducted using a 300-Ton AIDA servo press with a 25-Ton servo-cushion system. The punch speed was set at 15 stroke per minute (SPM), and a blank holder force of 50 KN were used. In the cross die forming test, the orientation of the blank can be changed to produce different forming behavior. Figure 1 shows the rolling direction (RD) of the blank, which is along the cross. Cross die forming test specimens are 470 mm x 470-mm rectangular sheets with thicknesses of 1mm for AA5052, 0.9mm for AA5754, and 1.2mm for AA6016. A stamping lubricant was applied on the both sides of the blank using the UNIST Motorized Roller System. The etched grid pattern was applied on each sample before the forming processes, and the strain distributions were measured by the grid analysis in ARGUS system, in which the strain values were calculated by comparing before and after grid shapes.

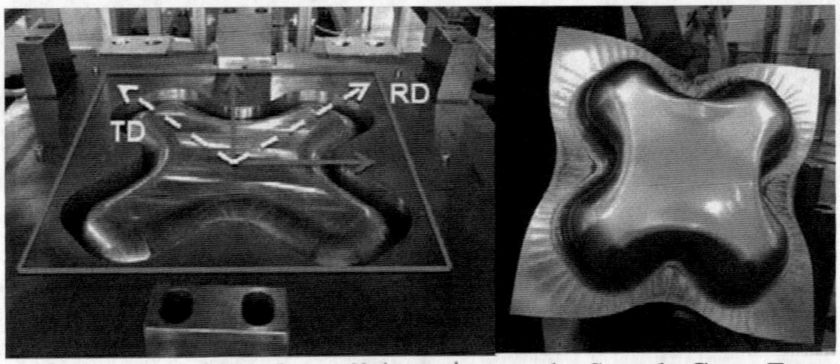

Figure 1. Schematic of Specimen Orientations and a Sample Cross Formed Part

Experimental results from the cross die forming tests were compared with the simulations result performed in LS Dyna explicit code using two anisotropic yield functions i.e. Hill's 48 and Yld2000-2D. It should be noted that for this study, Swift type hardening model was primarily selected over other types of hardening laws (e.g., Voce) because it exhibited the hardening behavior of the current material accurately. As for the simulations, element size of 5mm and a total number of 8836 quadrilateral shell elements were used for the blank. The blank was only constrained by friction in the analysis with no additional boundary conditions. The blank was placed between a blank-holder and a die. The blank-holder force was 50 KN. The punch was considered stationary, while the die moved toward the punch by 96 mm depth with a velocity of 300 mm/s. The friction coefficient of 0.15 was assumed. The thinning of the sheet is one of the most important factors to be considered in the sheet metal forming process. Figures 2-4 shows the thinning of Al5052 in depth of 90 mm, Al5057 in depth of 83 mm, and Al6016 in depth of 71 mm as the formability of these aluminum alloys are different. The contours for the experimental measurement and simulation results with Yld2000-2D are almost identical, which means this model is a better representative of the material behavior. It should be noted that the thinning measurement was not captured at the wall areas due to the removal of etched patterns of the aluminum surface. In Figure 5 cross formed samples were cut along two orientations i.e. along the cross and 45 degrees to the cross using a water-jet machining for thickness measurements which was done at 15mm intervals. The same procedure was performed in the simulations, in order to measure the thickness reduction of the deformed samples along the same orientations.

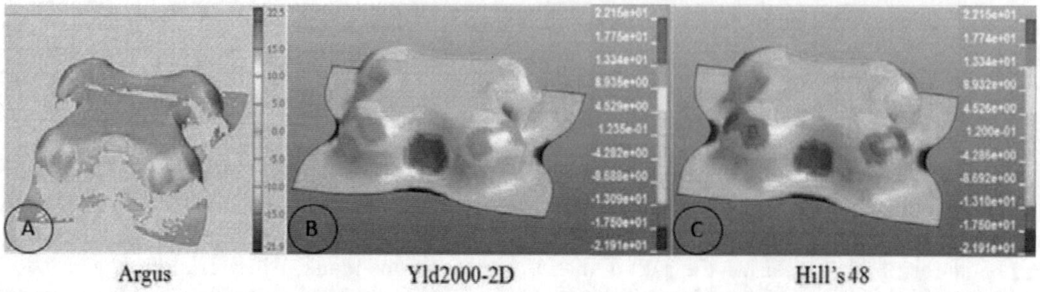

Figure 2. AA5052 thinning distribution contours: A) experimental measurement using Argus, B) FEM simulation results by using Yld2000-2D, C)FEM simulation results by using Hill's.

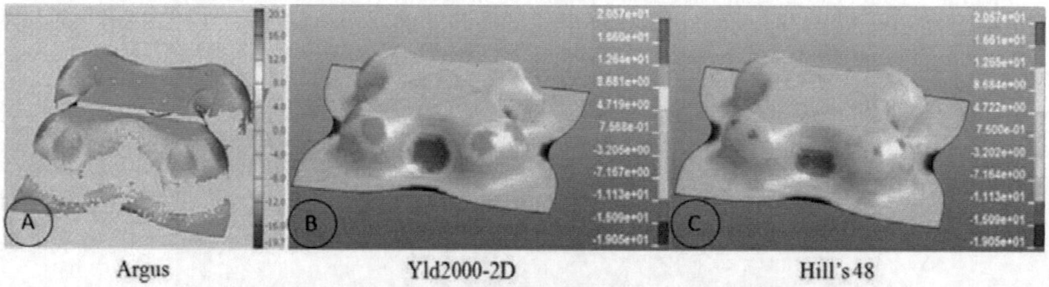

Figure 3. AA5754 thinning distribution contours: A) experimental measurement using Argus, B) FEM simulation results by using Yld2000-2D, C) FEM simulation results by using Hill's

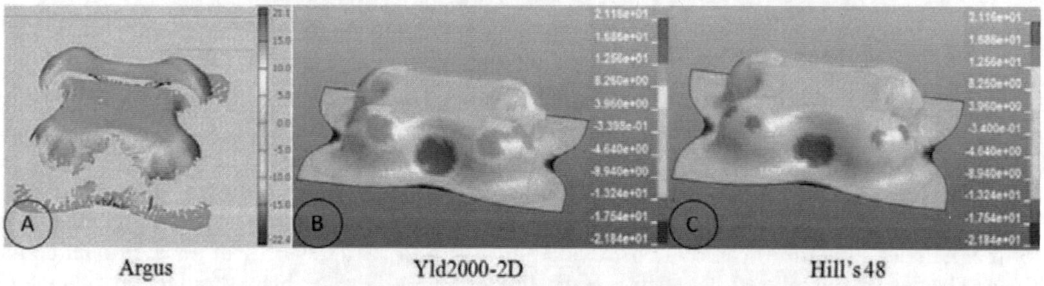

Figure 4. AA6016 thinning distribution contours: A) experimental measurement using Argus, B) FEM simulation results by using Yld2000-2D, C) FEM simulation results by using Hill's

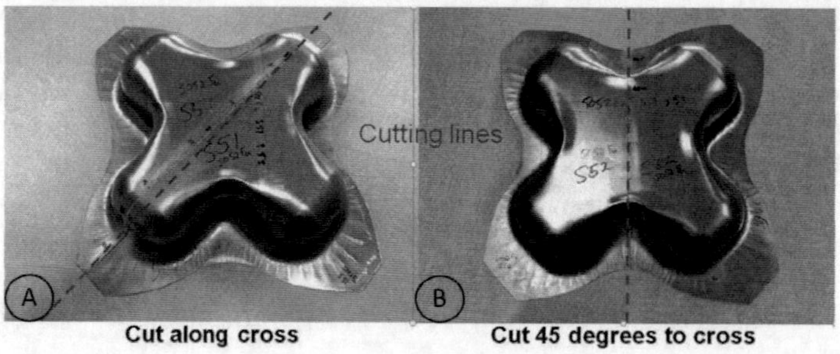

Figure 5. samples cut orientations for thickness measurements: A) along cross, B) 45 degrees to cross

4. Conclusions

This study showed that Yld2000-2D and even Hill's 48 are decent representative of anisotropy behavior of aluminum alloys namely; AA5052, AA5754 and AA6016 T-4 in the cross forming process. As for material characterization, uniaxial tensile tests of AA5052, AA5754 and AA6016 T-4 dog-bone specimen were performed to obtain the stress-strain curves and the anisotropic properties. Since a two-dimensional (2D) plane stress yield function is sufficient to capture the mechanical behavior in conventional metal forming processes, such as stamping, Yld2000-2D and Hill's 48 yield functions anisotropy coefficients were calibrated to predict the anisotropic properties of these aluminum sheets. Compared to Hill's 48, Yld2000-2D showed better prediction of the anisotropy (for both R-value and yield stress distribution). These two yield functions were used to simulate the cross die forming process in LS-Dyna explicit code and compared with the experiments conducted using a high capacity servo press with servo-cushion system. Furthermore, the thinning and strain distribution of the cross formed blanks obtained from FEM analysis were compared with the measured values obtained through ARGUS grid analysis. The simulation results and the experiments were almost identical especially with the Yld2000-2D.

Acknowledgement

This study was supported by EWI's Joint Industry Project (JIP) sponsors and the Ohio Development Service Agency (ODSA) on Advanced Manufacturing Program (AMP 15-08). The authors would like to thank the Forming Center of EWI for the support given to this research.

References

[1] Hirsch J 2017 Recent development in aluminium for automotive applications *Trans. Nonferrous Met. Soc. China* **24** 1995–2002

[2] Asgharzadeh A, Jamshidi Aval H and Serajzadeh S 2016 A Study on Flow Behavior of AA5086 Over a Wide Range of Temperatures *J. Mater. Eng. Perform.* **25** 1076–84

[3] Ebrahimi M, Djavanroodi F, Tiji S, Gholipour H and Gode C 2015 Experimental Investigation of the Equal Channel Forward Extrusion Process *Metals (Basel).* **5** 471–83

[4] Klos A, Kahrimanidis A, Wortberg D and Merklein M 2017 Experimental and numerical studies on the forming behavior of high strain Al-Mg-Si (-Cu) sheet alloys *Procedia Eng.* **183** 95–100

[5] Garware M, Kridli G T, Mallick P K, Garware M, Kridli G T, Mallick P K and Materials L A 2010 Tensile and Fatigue Behavior of Friction-Stir Welded Tailor-Welded Blank of Aluminum Alloy 5754 **19** 1161–71

[6] Kim H, Park T, Esmaeilpour R and Pourboghrat F 2018 Numerical Study of Incremental Sheet Forming Processes *J. Physics: Conf. Series.* **1063** 012017

[7] Laboratories A 1992 Finite element simulation of sheet metal forming for planar anisotropic metals **8** 453–76

[8] Hill R. A theory of the yielding and plastic flow of anisotropic metals. *Proc. Roy. Soc.* 1948;**193**, 281–297.

[9] Hill R 1987 CONSTITUTIVE DUAL POTENTIALS IN CLASSICAL PLASTICITY t / J (q) t / J (q) **35** 23–33

[10] Met J and Britain G 1993 A GENERAL YIELD CRITERTON USING BOUNDS AND A TRANSFORMATION WEIGHTING **41** 1859–86

[11] Bron F and Besson J 2004 A yield function for anisotropic materials Application to aluminum alloys **20** 937–63

[12] Pourboghrat F, Park T, Kim H, Mohammed B, Esmaeilpour R and Hector Jr L G, 2018 An integrated computational materials engineering approach for constitutive modelling of 3rd generation advanced high strength steels *J. Physics: Conf. Series.* **1063** 012010

[13] Park T, Kim H, Pourboghrat F and Esmaeilpour R 2018 A multi-scale modelling of 3rd generation advanced high strength steels to account for anisotropic evolution of yield surface and plastic potential *J. Physics: Conf. Series.* **1063** 012019

[14] Esmaeilpour R, Kim H, Park T, Pourboghrat F and Mohammed B 2017 Comparison of 3D yield functions for finite element simulation of single point incremental forming (SPIF) of aluminum 7075 *Int. J. Mech. Sci.* **133** 544–54

[15] Esmaeilpour R, Kim H, Park T, Pourboghrat F, Xu Z, Mohammed B and Abu-Farha F 2018 Calibration of Barlat Yld2004-18P Yield Function Using CPFEM and 3D RVE for the Simulation of Single Point Incremental Forming (SPIF) of 7075-O Aluminum Sheet *Int. J. Mech. Sci.* **145** 24–41

[16] Barlat F, Brem J C, Yoon J W, Chung K and Dick R E 2003 Plane stress yield function for aluminum alloy sheets — part 1 : theory **19** 1297–319

[17] Mohammed B, Park T, Pourboghrat F, Hu J, Esmaeilpour R and Abu-farha F 2017 Multiscale crystal plasticity modeling of multiphase advanced high strength steel strength steel *Int. J. Solids and Structures.* **000** 1–19

[18] Mohammed B, Park T, Kim H, Pourboghrat F and Esmaeilpour R 2018 The forming limit curve for multiphase advanced high strength steels based on crystal plasticity finite element modeling *Mater. Sci. Eng. A* **725** 250–66

IOP Conf. Series: Materials Science and Engineering **521** (2019) 012003 doi:10.1088/1757-899X/521/1/012003

Machinability of Alloy Steel and Titanium Alloy under Carbon Dioxide Snow, Micro-lubrication and Hybrid Lubro-Cooling

A Iqbal[1], J Zaini[1] and MM Nauman[1]

[1]Faculty of Integrated Technologies, Universiti Brunei Darussalam, Jln Tungku Link, Gadong BE1410, Brunei Darussalam

asif.asifiqbal@gmail.com

Abstract. Efficient removal of process heat in machining of high-strength alloys is critical to product quality, process cost and productivity. The paper experimentally investigates the effects of employing a cryogenic coolant (CO_2 snow), micro-lubrication and hybridization of the two on the performance measures, such as tool damage, machining forces and specific cutting energy in cutting of two commonly used high-strength alloys, Ti-6Al-4V and 40NiCrMo6. It was found that although the aforementioned cooling approached improved machinability of both the alloys but the titanium based alloy remained more difficult-to-cut than the alloy steel. Hybrid lubro-cooling and CO_2 snow cooling proved to be the most effective options for machining of 40NiCrMo6 and Ti-6Al-4V, respectively.

1. Introduction

Emulsion based flood coolant is in use for decades for dissipation of heat in machining of high-strength alloys but it has some limitations, such as limited cooling capacity, requirement of filtering and pumping, swarf generation and toxicity [1]. Due to extremely low operational temperatures, the cryogenic fluids possess capability of dissipating intense levels of process heat. Furthermore, they do not cause any waste or filtering related problems [2]. Micro-lubrication, also called as minimum quantity of lubrication (MQL), has also been utilized for enhancing machinability[3].

Dhananchezian et al have studied effects of applying liquid nitrogen (LN_2) jet at rake and flank faces through holes provided in the turning insert during the cutting of Ti–6Al–4V [4]. In another work, considerable improvement in tool life, by subduing adhesion, dissolution, and diffusion wear of tool, was reported with the use of LN_2 gas [5]. Improved surface integrity in the machining of a hardened steel is reported when cutting was done under cryogenic environment using PcBN inserts [6]. In another work, CO_2 snow was used in the machining of a high-strength β-titanium alloy (Ti-10V-2Fe-3Al) [7]. The authors have claimed substantial improvement in tool life, mainly due to subdual of notch wear. It was reported that under MQL environment, the major damage modes of coated carbide cutters in milling of an ultra-hard tool steel were suppressed [8]. Hwang et al have reported reduction in cutting forces and work surface roughness in machining of AISI 1045 steel under MQL [9]. It is reported that application of MQL on the tool's rake face in grooving of a β-titanium alloy leads to reduction in tool damage and deflections [10]. Sun et al have compared the effects of using flood coolant and MQL with cryogenic cooling in machining of an $\alpha + \beta$ titanium alloy [11].

In the context, the current work aims to investigate the effects of using CO_2 snow, micro-lubrication and their hybridization on machinability of a commonly used high-strength low-alloy steel (40NiCrMo6) and an $\alpha + \beta$ titanium alloy (Ti-6Al-4V).

(a) (b)

Figure 1. Experimental setup: (a) alloy steel work rod, MQL duct and CO2 carrying copper tubes; (b) titanium alloy work rod, CO2 carrying copper tubes, and accumulation of CO2 snow in the cutting region.

2. Experimental work

2.1. Predictors

The following predictors were controlled in the experimental runs of a cylindrical grooving process: (a) Work material (40NiCrMo6 and Ti-6Al-4V); (b) Work material's yield strength, R_p (870 MPa and 1,110 MPa); (c) Cooling mode: (1) No coolant (Dry), (2) CO_2 snow on the flank face only (FF-CO_2), (3) CO_2 snow on the rake face only (RF-CO_2), (4) CO_2 snow on the flank face and the rake face simultaneously (FF+RF-CO_2), (5) CO_2 snow on the flank face and MQL on the rake face simultaneously (FF-CO_2, RF-MQL), and (6) CO_2 snow on the rake face and MQL on the flank face simultaneously (RF-CO_2, FF-MQL); and (d) Cutting speed, V_c (80 m/min and 150 m/min).

A multiple-level full-factorial design of experiments was developed that generated 48 runs. The grooving inserts employed were TiN coated tungsten carbide having clearance angle, rake angle and cutting edge's width equal to 12.5°, 1.5°, and 3 mm, respectively. The work material rods were cut into diameters and lengths of 100 mm and 450 mm, respectively. A constant tool's feed rate of 0.12 mm/rev was used in all the experimental runs.

2.2. The setup

All the experimental runs were performed on a CNC lathe having a maximum motor power of 11kW. The micro-lubrication was supplied to the insert's face through a duct by forcing a vegetable oil at a flow rate of 25 mL/hr into the flow of air compressed at 6 bars of pressure. For the application of CO_2 snow at the insert's faces, copper tubes of 1 mm inner diameter were used for transportation of CO_2 gas from a bottle containing the gas. As the compressed gas comes out of the tube, it expands, converts to snow, and produces cooling effect in the surroundings in accordance with the Joule-Thomson phenomenon. Figure 1 presents the experimental setup.

2.3. Responses

The following responses were measured/determined for each experimental run: (a) Maximum width of the flank wear land of the insert after cutting a groove, *VB* (mm); (b) Specific cutting energy, *SCE* (J/mm^3); (c) Average values of the static and dynamic components of: (i) cutting force, F_{c_stat} (N) and F_{c_dyn} (N) and (ii) feed force, F_{c_stat} (N) and F_{c_dyn} (N).

Material removal rate, *MRR* (mm^3/sec), is a productivity measure, which is calculated using the formula, $MRR = 3 \times 0.12 \times (1000 \times V_c/60)$. The numbers "3" and "0.12" are the values of width of cut and feed rate, respectively. The average power consumed by each run was determined using a power clamp meter, Fluke 345. The *SCE* was obtained by dividing the average cutting power by the corresponding *MRR*. The flank wear was evaluated using a 10× toolmaker's microscope. The machining forces were measured by Kistler Piezoelectric dynamometer 9257B.

Table 1. Experimental results for the 48 runs

No.	Predictors				Responses					
	Work material	R_p (MPa)	Cooling mode	V_c (m/min)	VB (mm)	SCE (J/mm^3)	F_{c_stat} (N)	F_{c_dyn} (N)	F_{f_stat} (N)	F_{f_dyn} (N)
1	40NiCrMo6	870	Dry	80	0.064	2.744	803	439	524	1009
2	40NiCrMo6	870	Dry	150	0.078	2.218	939	504	606	1032
3	40NiCrMo6	870	FF-CO_2	80	0.049	2.912	803	313	553	887
4	40NiCrMo6	870	FF-CO_2	150	0.060	2.345	982	297	587	974
5	40NiCrMo6	870	RF-CO_2	80	0.051	2.438	750	416	594	961
6	40NiCrMo6	870	RF-CO_2	150	0.066	2.038	824	539	638	1042
7	40NiCrMo6	870	FF+RF-CO_2	80	0.044	2.781	739	457	564	966
8	40NiCrMo6	870	FF+RF-CO_2	150	0.051	2.130	893	515	555	1015
9	40NiCrMo6	870	FF-CO_2, RF-MQL	80	0.039	2.481	679	344	477	657
10	40NiCrMo6	870	FF-CO_2, RF-MQL	150	0.047	2.274	808	361	526	711
11	40NiCrMo6	870	RF-CO_2, FF-MQL	80	0.040	2.388	788	522	601	951
12	40NiCrMo6	870	RF-CO_2, FF-MQL	150	0.054	2.061	897	717	582	993
13	40NiCrMo6	1,100	Dry	80	0.097	3.485	1038	567	663	1132
14	40NiCrMo6	1,100	Dry	150	0.123	2.868	1185	589	698	1241
15	40NiCrMo6	1,100	FF-CO_2	80	0.057	3.487	1021	468	656	1064
16	40NiCrMo6	1,100	FF-CO_2	150	0.071	2.889	1217	438	707	1188
17	40NiCrMo6	1,100	RF-CO_2	80	0.061	2.913	862	624	662	1153
18	40NiCrMo6	1,100	RF-CO_2	150	0.083	2.413	1023	705	691	1286
19	40NiCrMo6	1,100	FF+RF-CO_2	80	0.060	3.344	970	672	677	1154
20	40NiCrMo6	1,100	FF+RF-CO_2	150	0.067	2.645	1106	613	634	1224
21	40NiCrMo6	1,100	FF-CO_2, RF-MQL	80	0.052	2.808	843	568	588	784
22	40NiCrMo6	1,100	FF-CO_2, RF-MQL	150	0.066	2.224	1020	633	630	953
23	40NiCrMo6	1,100	RF-CO_2, FF-MQL	80	0.055	3.065	933	824	776	1138
24	40NiCrMo6	1,100	RF-CO_2, FF-MQL	150	0.072	2.607	1113	868	814	1227
25	Ti-6Al-4V	870	Dry	80	0.108	3.513	748	301	587	419
26	Ti-6Al-4V	870	Dry	150	0.123	3.127	921	383	594	453
27	Ti-6Al-4V	870	FF-CO_2	80	0.072	3.819	694	267	581	559
28	Ti-6Al-4V	870	FF-CO_2	150	0.093	3.315	884	323	581	631
29	Ti-6Al-4V	870	RF-CO_2	80	0.09	3.367	638	488	695	914
30	Ti-6Al-4V	870	RF-CO_2	150	0.108	3.031	801	549	657	987
31	Ti-6Al-4V	870	FF+RF-CO_2	80	0.068	3.062	692	520	558	936
32	Ti-6Al-4V	870	FF+RF-CO_2	150	0.074	2.296	764	543	655	1000
33	Ti-6Al-4V	870	FF-CO_2, RF-MQL	80	0.071	3.571	665	396	501	380
34	Ti-6Al-4V	870	FF-CO_2, RF-MQL	150	0.093	3.280	742	470	615	456
35	Ti-6Al-4V	870	RF-CO_2, FF-MQL	80	0.103	3.010	655	668	637	858
36	Ti-6Al-4V	870	RF-CO_2, FF-MQL	150	0.118	2.414	708	713	675	932
37	Ti-6Al-4V	1,100	Dry	80	0.122	4.429	1223	423	683	603
38	Ti-6Al-4V	1,100	Dry	150	0.139	3.757	1416	535	684	1014
39	Ti-6Al-4V	1,100	FF-CO_2	80	0.075	4.983	1100	588	689	1422
40	Ti-6Al-4V	1,100	FF-CO_2	150	0.099	4.145	1173	644	827	1647
41	Ti-6Al-4V	1,100	RF-CO_2	80	0.105	4.060	903	944	788	2086
42	Ti-6Al-4V	1,100	RF-CO_2	150	0.122	3.595	988	1176	663	2638
43	Ti-6Al-4V	1,100	FF+RF-CO_2	80	0.072	3.806	801	1117	731	2196
44	Ti-6Al-4V	1,100	FF+RF-CO_2	150	0.088	2.839	883	1212	704	2348
45	Ti-6Al-4V	1,100	FF-CO_2, RF-MQL	80	0.075	4.506	1034	588	617	537
46	Ti-6Al-4V	1,100	FF-CO_2, RF-MQL	150	0.105	3.925	1212	651	611	856
47	Ti-6Al-4V	1,100	RF-CO_2, FF-MQL	80	0.109	3.744	869	1008	799	2003
48	Ti-6Al-4V	1,100	RF-CO_2, FF-MQL	150	0.123	3.021	1011	1180	969	2183

3. Experimental results

Table 1 presents the data obtained in respect of the aforementioned response variables. Analysis of Variance (ANOVA) was applied for quantification of the effects of the predictors and their interactions on all the responses. Table 2 presents the ANOVA results in respect of *VB* and *SCE*.

Table 2. Results of ANOVA applied to the data related to *VB* and *SCE*.

Source	Sum of Squares		DoF	Mean Squares		F-value		p-value	
	VB	*SCE*		*VB*	*SCE*	*VB*	*SCE*	*VB*	*SCE*
Work material	0.015	9.24	1	0.015	9.24	872	881	<0.0001	<0.0001
R_p	0.0023	4.65	1	0.0023	4.65	135	443	<0.0001	<0.0001
Cooling mode	0.0094	2.76	5	0.0019	0.55	109	53	<0.0001	<0.0001
V_c	0.0031	3.66	1	0.003	3.66	179	349	<0.0001	<0.0001
Work × R_p	0.0002	0.2	1	0.0002	0.2	14	19	0.0011	0.0003
Work × Cooling	0.0002	1.7	5	0.0003	0.34	20	32	<0.0001	<0.0001

3.1. Tool wear

The ANOVA results in respect of *VB* suggest that work material is the most influential parameter with regard to the tool wear. It's F-value (872) is way too high as compared to those of the other predictors. The titanium alloy is found to be a much more difficult-to-cut material as compared to the high-strength low alloy steel. It suggests that yield strength is not the only work material related parameter that makes a material difficult-to-cut. Beyond material strength, Ti-6Al-4V is also known to have high chemical affinity with various tool materials and low chip-rake contact length. The high levels of yield strength and cutting speed were found to significantly increase the tool wear. The effect of using various cooling modes on tool wear was also found to be significant, which suggests that CO_2 snow based cryogenic cooling possesses potentials to alter the way a cutting tool wears.

Numerical optimization was performed to find out the best cooling option for each of the two work materials for the sake of minimizing tool wear. The optimal cooling modes for cutting 40NiCrMo6 and Ti-6Al-4V were found to be FF-CO_2, RF-MQL and FF+RF-CO_2, respectively. The results suggest that application of CO_2 snow on the tool's flank face causes reduction in the magnitude of flank wear regardless of the work material used. Super-cool jet of CO_2 effectively removes accumulated heat from the cutting edge and vicinity, thereby, suppressing temperature dependent tool wear modes. MQL has shown promising effect only when applied onto the rake face. The *Work × Cooling* interaction suggests that the effect of work material on *VB* is most influential when a combination of CO_2 snow and micro-lubrication is used on the opposite tool faces.

The modes of tool wear were analyzed using optical micrography. Two modes of wear, abrasion and wear with various magnitudes of intensity were clearly observed. A complete analysis of the photomicrographs revealed that the adhesive mode of wear was observed in machining of the titanium alloy with its intensity in proportion with the material's yield strength and cutting speed. On the other hand, the intensity of abrasive wear was not dependent upon work material but was in proportion with material's yield strength. Figure 2 presents the micrographs of the selected cutting edges.

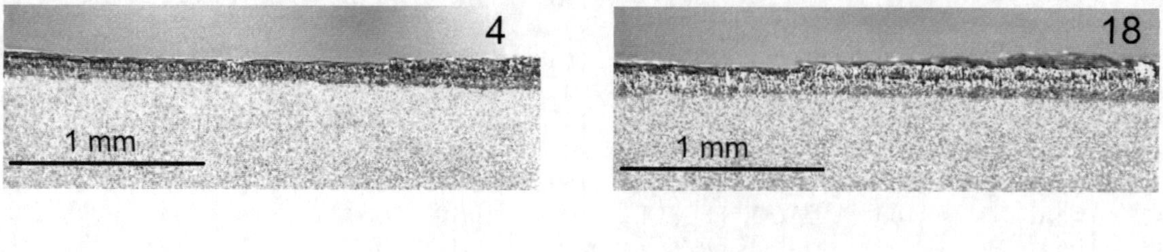

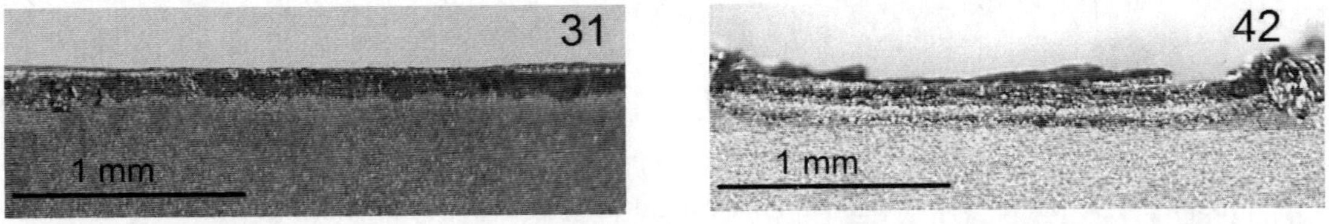

Figure 2. Optical micrographs of the inserts used in the experiments number 4, 18, 31 and 42.

3.2. Specific cutting energy

The ANOVA results in respect of *SCE,* as presented in Table 2, suggest that work material tops the list with respect to strength of effect followed by yield strength, cutting speed and cooling mode. An equally strong Ti-6Al-4V causes more consumption of cutting energy than does the alloy steel because high strength causes accelerated damage to the tool. Cutting with damaged tool geometry demands a high level of cutting force to overcome yield strength of the material, which causes an increase in cutting energy required to cut a unit volume of material. Furthermore, stronger temper of material is cut with a higher level of cutting energy. Cutting speed also has a significant effect on *SCE* with its high level causing reduction in the response's value. Moreover, almost all the cooling modes involving CO_2 snow seem to have a favorable effect on *SCE*. Like tool wear, the optimal cooling modes for the sake of minimizing *SCE* in cutting of 40NiCrMo6 and Ti-6Al-4V are FF-CO_2, RF-MQL and FF+RF-CO_2, respectively. The interaction *Work × Cooling* exerts a strong influence on *SCE*. A deeper analysis revealed that the effect of work material's choice is most significant when the applied cooling mode is FF-CO_2, RF-MQL and is least significant when CO_2 is applied on both the faces simultaneously.

3.3. Machining forces

Exceedingly high levels of machining forces, especially in cutting of the stronger temper of the titanium alloy, were observed, as shown in Table 1. ANOVA was applied on the data related to the components of the machining forces. Yield strength was found to be the most influential predictor for all the four force components. Large machining forces were needed to generate the stresses required to overcome the increased level of strength in cutting of the stronger tempers of both the work materials. For most of the cooling modes employed, yield strength has a much more influential effect on the force component when the work material is Ti-6Al-4V.

4. Conclusions

The paper presents an experimental investigation for machinability comparison of Ti-6Al-4V and 40NiCrMo6 under cryogenic and hybrid lubro-cooling environments. Ti-6Al-4V was found to possess poorer machinability than an equally strong high-strength low alloy steel. Though, most of the cryogenic and hybrid lubro-cooling techniques improve the machinability characteristics of both the alloys but the titanium alloy still remains more difficult to cut. Application of CO_2 snow at the flank face of a tool was found to be beneficial for the machinability of both the alloys. Having applied CO_2 snow at the flank face, the option of using micro-lubrication at the rake face proved to be the most suitable one for the machining of 40NiCrMo6. However, the option of applying CO_2 snow on both the tool faces proved to be the most beneficial one for Ti-6Al-4V.

5. References

[1] De Lacalle LL, Angulo C, Lamikiz A and Sanchez JA 2006 *J. Mat. Proc. Technol.* **172** 11
[2] Hong SY 2001 *J. Manuf. Sci. & Eng.* **123** 331
[3] Park KH, Olortegui-Yume J, Yoon MC, and Kwon P 2010 *Int. J Mach. Tools & Manuf.* **50** 824
[4] Dhananchezian M and Kumar MP 2011 *Cryogenics* **51** 34
[5] Venugopal KA, Paul S and Chattopadhyay AB 2007 *Wear* **262** 1071
[6] Umbrello D, Micari F and Jawahir IS 2012 *CIRP Ann. Manuf. Technol.* **61** 103
[7] Machai C and Biermann D 2011 *J. Mat. Proc. Technol.* **211** 1175

[8] Iqbal A, Dar NU, He N, Khan I and Li L. 2009. *Proc. Inst. Mech. Eng. Part B J. Eng. Manuf.* **223** 43.
[9] Hwang YK and Lee CM 2010 J. *Mech. Sci. & Technol.* **24** 1669
[10] Iqbal A, Biermann D, Abbas H, Al-Ghamdi KA and Metzger M 2018 *Int. J. Adv. Manuf. Technol.* DOI: 10.1007/s00170-018-2267-4
[11] Sun Y, Huang B, Puleo DA and Jawahir IS 2015 *Procedia CIRP* **31** 477

IOP Conf. Series: Materials Science and Engineering **521** (2019) 012004 doi:10.1088/1757-899X/521/1/012004

Thermomechanical Study and Fracture Properties of Silicon Wafer under Effect of Crystal Orientation

Mohamed Cherif Ben Romdhane[1], Hatem Mrad[1,*], Fouad Erchiqui[1], Ridha Ben Mrad[2]

[1]École de génie, Université du Québec en Abitibi-Témiscamingue, 445 Boulevard de l'Université, Rouyn-Noranda, QC J9X 5E4, Canada.
[2]University of Toronto, 5 King's College Rd., Toronto, Ontario, M5S 3G8, Canada.

*Corresponding author: hatem.mrad@uqat.ca

Abstract. Designing and optimizing the performance of a piezoresistive pressure microsensor involves not only studying the piezoresistive properties of polysilicon but also analyse the thermomechanical behaviour of its most sensitive internal part. This work focuses on an analytical, numerical and experimental study of a silicon wafer used in a pressure sensor. A thermomechanical study was conducted to assess the effects of temperature and material orientation on the deflexion and failure mode of Silicon membrane. The chosen analytical approach is based on Kirchhoff theory in which a square Silicon membrane is subjected to a distributed pressure over all its surface. Thermomechanical simulations and fracture analysis are carried out using Abaqus software. An experimental thermomechanical technique to determine maximum wafer deflexion and its effect on crack initiation and fracture mode was implemented. Deflexions tests were carried out on a bending machine using specimens of p-Si doped wafers. Deflexion values are measured for each applied temperature. Some preliminary tests were performed to determine the impact of temperature and Silicon wafers orientation on maximum deflexion when fracture occurs. The test's results confirmed that the failure mode is highly brittle and follow each time crystal orientations 0° and 45°. Furthermore, at room temperature, fracture occur at 389 µm of displacement, and grows by 5.2 % when temperature is increased over 40°C. Consequently, Silicon sensitivity will be increased by 4%.

1. Introduction

The use of microsystems such as pressure sensors has increased due to their small size, high sensitivity and accuracy. Basically, there are two main classes of pressure sensors: piezoresistive and capacitive, but the principle of pressure detection has remained unchanged. To get the best design and quality of those devices, it shall simulate their behaviour to reduce additional development costs, its sensitivity and failure risks. The following section present a summary of the functioning principle of a Silicon piezoresistive pressure sensor with a brief theory background. The temperature dependency of thermal and mechanical properties such as expansion, conduction coefficients and Young's modulus in the interval 20°C to 160°C are identified based previous publications [1] [2] [3]. A piezoresistive pressure sensor with movable diaphragm is composed mainly of an oxidized Silicon membrane covered with oxide as shown in figure 1. When a pressure is applied on the diaphragm, four piezoresistive polysilicon gauges placed over its top surface convert each mechanical deformation of the diaphragm into resistance variation. Sensor output is obtained by arranging gauges to form a Wheatstone bridge [4]. For designing and simulating the sensors, various aspects, such as fabrication technology, diaphragm thickness and orthotropic properties of silicon are considered. The applied pressure causes diaphragm deflection and stress to be measured by piezoresistors often placed in regions where high stress is generated. The

IOP Conf. Series: Materials Science and Engineering **521** (2019) 012004 doi:10.1088/1757-899X/521/1/012004

diaphragm thickness, shape, material orientation and mechanical properties as well as the piezoresistors orientation and location are important factors which affect the measured pressure.

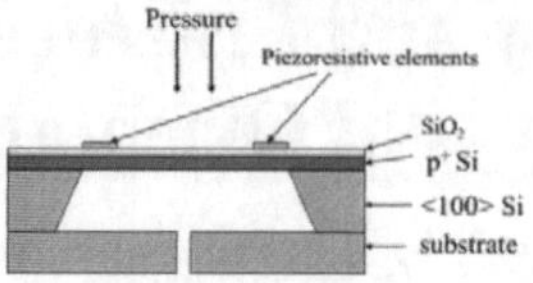

Figure 1. Sectional drawing of a piezoresistive silicon-on-insulator membrane sensor.

The change in resistance of a piezoresistive material under stress is obtained by following equation:

$$\frac{\Delta R}{R}=\frac{\Delta L}{L}-\frac{\Delta A}{A}+\frac{\Delta \rho}{\rho} \quad (1)$$

Where $\Delta L/L$ and $\Delta A/A$ are geometrical deformation regarding length and cross section of the resistors, respectively and ρ is the resistivity.

These terms are negligible compared to $\Delta\rho/\rho$ in case of piezoresistors under stress [5]. Hence, the equation can be rewritten as follows:

$$\frac{\Delta R}{R}=\frac{\Delta \rho}{\rho}=\pi_l\sigma_l+\pi_t\sigma_t \quad (2)$$

Where, π_l and π_t is longitudinal and transverse piezoresistive coefficients. σ_l and σ_t are the longitudinal and transverse stresses acting on the piezoresistor.

The piezoresistors are oriented along [110] directions on (100) wafer to maximize the piezoresistive effect. In Table 1, we present the piezoresistive coefficients for the transversal and longitudinal configurations of the gauges.

Table 1. Piezoresistive coefficients (p-Si doping types) for the transversal and longitudinal configurations of the gauges.

Stresses direction	Current direction	Configuration	Piezoresistives coefficients
<100>	<100>	Longitudinal	π_{11}
<100>	<010>	Transversal	π_{12}
<110>	<110>	Longitudinal	$(\pi_{11}+\pi_{12}+\pi_{44})/2$
<110>	<1-1 0>	Transversal	$(\pi_{11}+\pi_{12}-\pi_{44})/2$
<111>	<111>	Longitudinal	$(\pi_{11}+2\pi_{12}+2\pi_{44})/2$

Piezoresistive coefficient	p-Si (resistivity =7.8Ω cm)
π_{11}	6.6
π_{12}	-1.1
π_{44}	138.1

2. Results and Discussion

2.1. Thermomechanical fracture simulation

A thermomechanical (TM) model of Silicon wafer fracture was developed using the finite element analysis (FEA). Finite element model of a Silicon wafer (diameter =50 mm, thickness=279 μm, crystal direction=100) is shown in figure 2-a. The central square surface is left free to apply pressure or concentrate force using a punch with a maximum descent speed of 0.5 mm/s. Initial temperature applied over the entire wafer is equivalent to the room's temperature. The punch is a rigid body and its descent speed corresponds to that used in experimental tests. The Silicon wafer is held between two superposed male-female supports in order to leave free only the central square surface. Figure 2-b shows the ideal configuration of contact punch/wafer to analyse maximum deflexion propagation, fracture and stress distribution at various fixed temperatures. The test's results confirmed that the failure mode is highly brittle and follow each time crystal orientations 0° and 45° [6]. These outputs are determined both by experimental tests and numerical simulation using FEA.

IOP Conf. Series: Materials Science and Engineering **521** (2019) 012004 doi:10.1088/1757-899X/521/1/012004

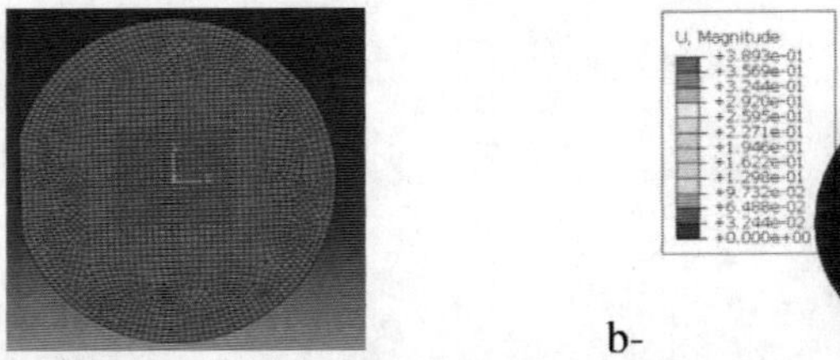

a - b-

Figure 2. Finite element model of: a- the Silicon wafer;b- Contact configuration wafer/punch.

Preliminary tests indicate that fracture start at limit deflexion equal to 389 μm for a temperature of 100 °C. Finite element fracture simulations using a flat bottom punch confirmed the similar trends as shown in figure 3. To illustrate fracture propagation process, the punch course exceeds the limit requests of Silicon wafer. During the simulation tests, two wafers orientations are considered 0° and 45° according to direction <110>. Between these two orientations, mechanical properties varies by up to 45%.

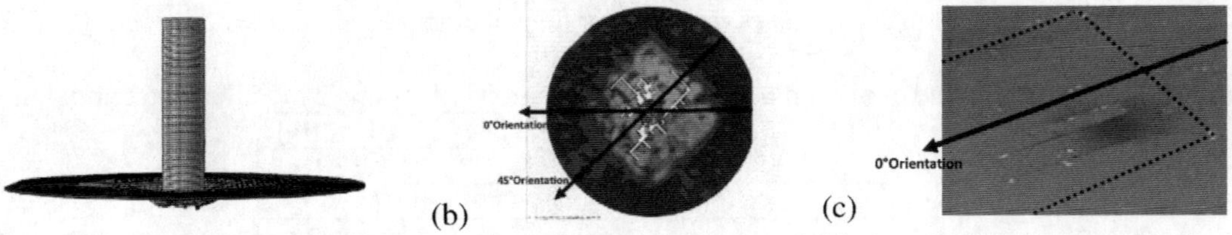

(a) (b) (c)

Figure 3. FEA simulation and experimental of fracture. a- Fracture initiation; b- FEA fractured wafer: material orientation 45°, c- Experimental fracture for wafer oriented at 0°.

Figure 4 shows curves of maximum deflection versus temperature 20°C to 200°C. Both experimental and thermomechanical FEA (TM-FEA) are reported when Young's modulus is equal to 130 GPa and 169 GPa respectively for chosen 0° and 45° orientations.

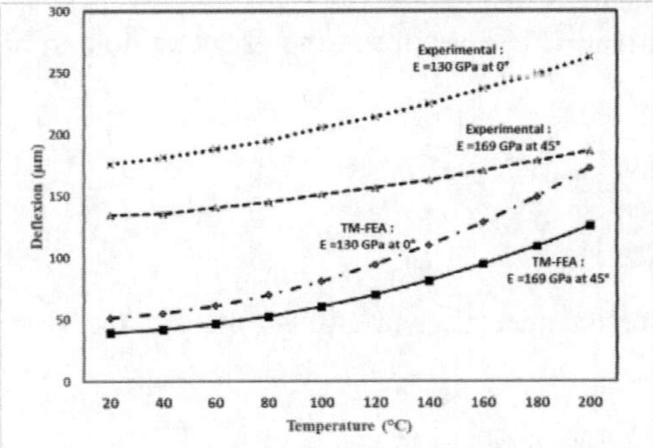

Figure 4. Variation of the deflection according to temperature.

2.2. *Cristal orientation*

The most common wafer type is called "(100) wafer", which means that the wafer's surface is parallel to the silicon plane. According to <110> crystal direction, figure 5c shows typical directions X and Y in layout (100) for MEMS design [7]. The sample is placed in a well-defined orientation (plane 110) to ensure the correct positioning on the support. Young's modulus and Poisson's coefficient varies from 130 GPa to 169 GPa and from 0.28 to 0.36 respectively for one standard orientation at 0° to the maximum one at 45°[8]. Figure 5a and b present the variation of Young's modulus and the Poisson's coefficient according to crystal orientation 0° to 90° [9].

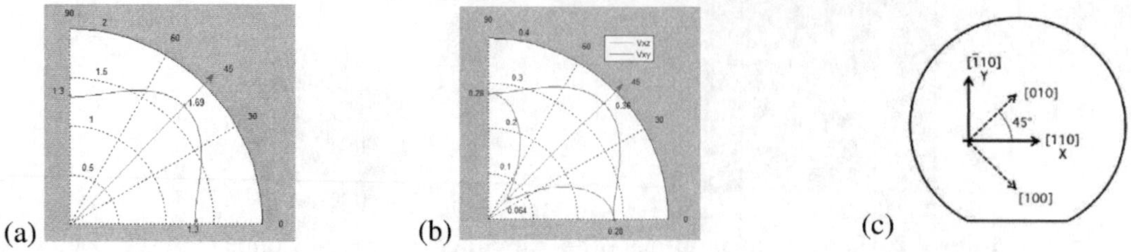

Figure 5. Mechanical properties variation at 0° and 45° crystal orientations. a- Young's modulus (10^{11}GPa); b- Poisson's ratio; c- Typical design orientation of Silicon wafers.

Most silicon wafers are not pure silicon; a certain amount of chemical impurities is usually added to control the electrical properties of the wafer. The volumetric effect of adding the mass of dopant atoms to the crystal lattice is negligible [10]. Since Young's modulus variations can reach 45%, it was decreased by 10%, 20% and 30% respectively with respect to its average value, while leaving constant a same applied load equal to 25N. The corresponding displacement evolution is described in Table 2.

Table 2. Displacement rate evolution depending on variations of Young's modulus and temperature.

	Young's modulus variations (%)		
	10	20	30
Temperature (°C)	Displacement (%)		
0	3.09	3.29	3.47
50	2.52	2.93	3.28
100	2.37	2.81	3.24
200	1.67	2.29	2.86
300	1.31	1.86	2.54

By decreasing Young's modulus values, displacement and stress will be increased consequently for the same applied load. However, increasing temperature generate opposite effect. It can be concluded that the direction of the plans must be respected to find the desired results.
The longitudinal piezoresistive coefficient in the equation at <110> is approximately equal to half of π_{44}, where π_{44} is the piezoresistive shear coefficient.

$$\frac{\Delta R_1}{R_1} = \frac{\Delta R_3}{R_3} = \pi_{l[110]}\sigma_{[110]} + \pi_{l[110]}\sigma_{[1\bar{1}0]}$$
$$= \frac{1}{2}\pi_{44}\left(\sigma_{[110]} - \sigma_{[1\bar{1}0]}\right) \approx \frac{1}{2}\pi_{44}\sigma_{[110]} \quad (3)$$

The two piezoresistors (R1 and R3) are the longitudinal resistances, the variation of these two resistances can be calculated as follows [11]:

$$\frac{\Delta R_2}{R_2} = \frac{\Delta R_4}{R_4} = \pi_{l[110]}\sigma_{[110]} + \pi_{l[110]}\sigma_{[1\bar{1}0]}$$
$$= \frac{1}{2}\pi_{44}\left(\sigma_{[110]} - \sigma_{[1\bar{1}0]}\right) \approx -\frac{1}{2}\pi_{44}\sigma_{[1\bar{1}0]} \quad (4)$$

When the diaphragm is loaded, the longitudinal and transversal gauge resistors show opposite variations. The variation of the output voltage with the loaded pressure is as follows:

$$V_{out} = \frac{(R_1+\Delta R_1)(R_3+\Delta R_3)-(R_2+\Delta R_2)(R_4+\Delta R_4)}{R_1+\Delta R_1+R_4-\Delta R_4+R_3+\Delta R_3+R_2-\Delta R_2} \quad (5)$$

Where the design meets perfect manufacture qualities, then $R_1 = R_2 = R_3 = R_4 = R$ and $\Delta R_1 = \Delta R_2 = \Delta R_3 = \Delta R_4 = \Delta R$. Output voltage can be rewritten as:

$$V_{out} = I \times \Delta R = I \times \pi \times \sigma \times R \quad (6)$$

The sensitivity of a pressure sensor is defined as the relative change in the output voltage per unit of applied pressure [12]:

$$S = \frac{\Delta V_{out}}{\Delta P}\frac{1}{V_{in}} = \frac{\Delta R}{\Delta P}\cdot\frac{1}{R} \quad (7)$$

Table 3 shows sensitivity variation for a temperature of 5°C to 100°C.

Table 3. The sensitivity temperature interval 5-100 °C.

Temperature (°C)	Sensitivity (mV/V/MPa)
5	3.39
25	3.48
50	3.52
75	3.61
100	3.96

3. Conclusion

In this article we have studied the influence of temperature and silicon crystal orientation on its thermomechanical behaviour and the maximum deflexion of a wafer used on piezoresistive pressure sensors. Two parameters were evaluated: the displacement of the diaphragm under different temperatures and the sensor sensitivity.

From 25°C to 80°C, the displacement variation was of 7.72%. On the other hand, it is 28.41% between 80 and 200. This variation is the result of the evolution of the coefficient of thermal expansion which changes by more than 50% between 25 and 150°C. This variation is the result of the evolution of the coefficient of thermal expansion which changes by more than 50% between 25 and 150°C [2]. In this case the sensor will be functional below 100°C.

Silicon wafer sensitivity was increased by 17% between 5 and 100°C, 2.65% between 5 and 25°C, 3.75% between room temperature and 75°C. Crack size and starting are proportional to temperature variation. During room temperature experiments tests, fracture appeared for a displacement of 389 μm when a force of 25N was applied.

4. Référeces

[1] R. Hull, Properties of crystalline silicon: IET, 1999.

[2] C. Glassbrenner and G. A. Slack. Thermal conductivity of silicon and germanium from 3 K to the melting point. *Physical Review,* vol. 134, p. A1058, 1964.

[3] G. Holland and L. J. Neuringer. Proceedings of the International Conference on the Physics of Semiconductors, Exeter (The Institute of Physics and Physical Society, London, p. 474. 1962.

[4] A. Dewdney. Nanotechnology-Wherein molecular computers control tiny circulatory submarines. *Scientific American,* vol. 258, pp. 100-&, 1988.

[5] M. Bao. Analysis and Design Principles of MEMS Devices. *Elsevier,* 2005.

[6] G. P. Cherepanov. Mechanics of Brittle Fracture. 1979.

[7] B. Bhushan and X. Li. Micromechanical and tribological characterization of doped single-crystal silicon and polysilicon films for microelectromechanical systems devices. *Journal of Materials Research,* vol. 12, pp. 54-63, 1997.

[8] M. A. Hopcroft, W. D. Nix, and T. W. Kenny. What is the Young's Modulus of Silicon?. *Journal of microelectromechanical systems,* vol. 19, pp. 229-238, 2010.

[9] H. McSkimin, W. Bond, E. Buehler, and G. Teal. Measurement of the elastic constants of silicon single crystals and their thermal coefficients. *Physical Review,* vol. 83, p. 1080, 1951.

[10] B. Abeles, D. Beers, G. Cody, and J. Dismukes. Thermal conductivity of Ge-Si alloys at high temperatures. *Physical review,* vol. 125, p. 44, 1962.

[11] X. Li, Q. Liu, S. Pang, K. Xu, H. Tang, and C. Sun. High-temperature piezoresistive pressure sensor based on implantation of oxygen into silicon wafer. *Sensors and Actuators A: Physical,* vol. 179, pp. 277-282, 2012.

[12] M. Gad-el-Hak. The MEMS Handbook. 2002.

IOP Conf. Series: Materials Science and Engineering **521** (2019) 012005 doi:10.1088/1757-899X/521/1/012005

Obtaining Radiation-resistant Material by SHS Method

Maria Kozhanova[1], Anton Kozhanov[1], Igor Golubenko[2], Narkiza Bikberdina[3], Marina Boronenko[3, a)]

[1]Dimitrovgrad Engineering and Technological Institute of the National Research Nuclear University MEPhI, Dimitrovgad, Russia; [2]OOO Resonance Plus, Dimitrovgad, Russia; [3]Yugra State University, Khanty-Mansiysk, Russia;

[a)]Corresponding author: MarinaB7@yandex.ru

Abstract. All The first results of obtaining radiation-resistant material by the SHS method are presented in the article. Synthesis of material based on the system Fe2O3-Al-Cr2O3-MoO3-Ti. The resulting material was irradiated with an electron beam of 1.8 MeV. The synthesis process was filmed by an action video camera. Fourier image analysis revealed the instability of combustion regimes and predicted the quality of reaction products. A comparative analysis of the microstructure of the samples before and after irradiation with electrons confirmed the results of the FFT. Despite the relatively low quality of the synthesis products, the embrittlement occurred at the absorbed dose of 70 Mrad. In the future, it is planned to improve the methodology for identifying unstable synthesis modes and adjusting the percentage ratio of the charge components.

1. Introduction

The problem of providing radiation-resistant materials has been relevant for several decades, and now it will remain important. There are the following areas of technology in which such materials are necessary to ensure long-term operation of measuring instruments and structures. Self-propagating high-temperature synthesis (SHS) is an efficient, economical, energy-saving and energy-saving synthesis method for materials with desired properties [1-3]. The advantages of self-propagating high-temperature synthesis, compared with traditional methods, are speed, low power consumption, unique structure and properties of synthesis products. In the wave of synthesis, purification from harmful impurities occurs. Due to this, SHS products are distinguished by high purity. However, there are unsolved problems. You cannot get steel or alloy, which includes a large number of ingredients. A large number of metal ligaments and alloying additives, which are inert, worsens the exothermicity of the mixture. This leads to incomplete recovery of elements from oxides. Replacing alloying additives with finished metals makes the process uneconomical. At the same time, it is known that the size of the inert additive can be up to 30% by weight of the reactants. Thus, the relevance of research is maintained. We attempted to obtain a radiation-resistant material by the SHS method. The purpose of this article is to describe the first results.

Temperature and speed of processes affecting the properties of synthesis products. The method of monitoring the thermal imager allows not only to estimate the average speed and temperature of the reaction, but also to observe their dynamics in the area of interest, to monitor the movement and the evolution of regions from temperature. Therefore, the control of the main thermophysical parameters is carried out using the method of high-speed video.

IOP Conf. Series: Materials Science and Engineering **521** (2019) 012005 doi:10.1088/1757-899X/521/1/012005

2. Research Methodology

SHS cast refractory materials - a complex multi-step process. There are three basic steps: burning, phase separation and cooling of combustion products. Combustion takes place in two stages: the reconstitution of the elements of an oxide (metallothermy) and reacting the recovered elements. Prevalence of iron oxides cheapness and high calorie thermite mixture (8400kDzh / kg) allows its use for the synthesis. The composition of the mixture of powders must satisfy the equation [1-3]:

$$\left[\sum_{i=1}^{N_{MO_x}} \nu_i (MO_x)_i + Al\right] + \sum_{i=1}^{N_x} \nu_i X_i = \sum_{j=j}^{N_p} \nu_i P_i + Al_2O_3$$
,

where (MO_x)-oxides; X_i- metallic or non-metallic reagents; *Pi*-solid phase products, coefficients before exponential factor, ν_i- stoichiometric coefficients.

The adiabatic combustion temperature, chemical and phase composition of products depend on the ratio of reactants in the starting mixture, their initial temperature, the size of the reaction volume and the pressure of the surrounding gaseous medium. The composition of the equilibrium final products of synthesis is determined by minimizing the thermodynamic potential of the system. Vapors and suboxides released in the melt during combustion, lead to scatter of the melt. After passing through the combustion front, a two-phase melt is formed, in which the metal oxide – reducing agent (oxide phase) forms a continuous medium. Drops of the refractory compound of the reduced metal (metal phase) are distributed in it. Due to the difference in the densities of the phases under the action of the field of gravity, the mutual movement of the phases occurs. The motion occurs in the cooling melt under conditions of increasing viscosity of the oxide phase. Termination of phase separation occurs either at the time of complete release of drops into the "metal" ingot from the melt (full phase separation), or at the time of crystallization of the oxide phase (incomplete phase separation). Expression for the dependence of the completeness of phase separation on the main parameters [1-3]:

$$\eta_F = A\frac{d_k g}{\alpha\mu_0}\frac{r}{r + L_0}exp\left(-\frac{E_v}{RT_g}\right)\left[1 - exp\left(-\frac{E_v(T_g - T_F)}{RT_g^2}\right)\right],$$

where g is the acceleration of free fall; T_F is the temperature at which phase separation ends; T_g is the burning temperature, d_k is the droplet diameter; μ_0 is the pre-exponential viscosity of the oxide phase, E_v is the activation energy of a viscous flow, r is the radius of the sample, L_0 is the height of the sample, α is the heat transfer coefficient from the sample.

Obviously, the strongest parameter that allows you to adjust the completeness of phase separation η_F is the combustion temperature. Under conditions of intense cooling, complete melt separation requires substantial overheating of the melt above the melting point of the oxide phase. To maintain the required combustion temperature, the spatial and temporal uniformity of the emergence of new combustion centers is needed. If we supplement the optical part of the system with a microscope, we can measure the brightness temperature with a high spatial resolution. To study the process of phase separation, you can use optoelectronic micropyrometric systems. Trace-analysis [] and FFT video recording of the wave propagation process of SHS. Video filming converts the synthesis process into a sequence of images, each pixel (*x*, *y*) of which has a brightness proportional to the instantaneous temperature value of the corresponding element of the reactant:

$g(x, y, T) = T[f(x, y, I)]$, where $f(x, y, T)$ is the original object, $g(x, y, I)$ is the image, T is the transfer function of the optoelectronic system. Next, you need to find the interframe difference of the video sequence $\Delta X(t, y, I) = X(t, y, I) - X(t-\Delta t, y, I)$ and set the mask that limits the phase separation temperature.

The Fourier transform of a discrete function in one variable $f(x)$, x = 0, 1, 2, ..., M-1, is given by

$$F(u) = \frac{1}{M}\sum_{u=0}^{M-1} f(x)e^{-i2\pi ux/M}, x = 0,1,2,...M-1.$$

The power spectrum of the recorded signal can be estimated from the amplitude of the transformed Fourier data. Comparing the power spectra for the beginning, middle and end of the reaction, one can understand whether the concentration of alloying additives was critical.

IOP Conf. Series: Materials Science and Engineering **521** (2019) 012005 doi:10.1088/1757-899X/521/1/012005

3. Experimental equipment

To apply optical pyrometry, the radiation of a heated body must obey Kirchhoff's law. Usually solids and liquids when handling this requirement. Brightness pyrometry uses the equality of the brightness of the object under study and the brightness of an absolutely black body at the same wavelength λ_0. For calibration, we used the TRU 2350 temperature lamp as a reference. The radiation of real bodies is different from the level of a black body. The formula relating the brightness and true temperature of a real body through the spectral emissivity (monochromatic degree of blackness) $\varepsilon_{\lambda T}$:

$$\frac{1}{T} = \frac{1}{T_b} + \frac{\lambda_0}{c_2} \ln \varepsilon_{\lambda_0, T},$$

The where $c_2 = (1{,}43880 \pm 0{,}00019) \cdot 10^{-2}$ m · K

After calibration, you can use a video camera as a bright thermal imager. For a detailed study of the processes occurring in the wave of combustion of SHS [8-9], a microscope objective was used. The lens was mounted on a digital electronic video camera "ASI120MC Camera". The resolution of the video camera is 1.2 megapixels 1280 × 960, pixel size: 3.75 microns. The result is a digital microscope that is directly connected to a computer. Video recording is carried out at a speed of 30 frames per second. Working distance 100 mm. The digital electron microscope is used to measure speed and local temperature, the viewing area is 1.5x1.5 cm. The additional sports video camera "Soocoo S70" is designed to monitor the temperature and speed of the combustion process as a whole. On both video cameras, welding glasses were used as additional filters. A general view of the experimental setup, video footage of the combustion process, filmed «Soocoo S70», they are shown in figures 1 and 2 respectively.

Figure 1. These two figures have been placed side-by-side to save space. Justify the caption. experimental setup.

Figure 2. Video frame examples of a typical synthesis process taken by a sports camera «Soocoo S70».

The experiments were carried out using several variants of the elemental composition of the powder mixture. Here we present the results of the synthesis without carbon and nickel. For the synthesis, PA4 aluminum powder, molybdenum oxide and chromium oxide with three valence dispersions up to 50 microns, iron oxide with dispersions up to 300 microns and titanium powder up to 100 microns were used. The powders were mixed in mass percent: Fe_2O_3-57.77 wt.%; Al-30.13 wt.%, Cr_2O_3, 9.09 wt.%, MoO_3-1.05 wt.%, Ti-2.03 wt.%. The mixture of powders was placed in a quartz open-type reactor (test tube). The quartz reactor with the charge was mounted on a tripod, set so that the edges of the test tubes were flush with the refractory disk. Then all the smoke and sparks flew into the pipe. Combustion was initiated by the tungsten coil. After cooling, the synthesis products were separated from the slag. Next, prepare the thin sections. The microstructure of the samples was studied on an Axiovert-200 metallographic microscope using an image processing system. Structure VideoTest-5.

IOP Conf. Series: Materials Science and Engineering **521** (2019) 012005 doi:10.1088/1757-899X/521/1/012005

Each sample was divided into two parts. One part was sent to Dimitrovgrad for irradiation, the second was given for X-ray analysis. Irradiation was performed on a pulsed linear accelerator ILU-6. Irradiation occurred at room temperature; pulse frequency - 12.5 Hz; current 0.39A; electron energy 1.8 MeV in an atmosphere of oxygen and ozone; irradiated area 20 × 10 cm. X-ray diffraction analysis was performed on an ARL X`TRA diffractometer.

4. Results and discussion

Experience shows that when a combustion wave moves in a sample volume, a substance is transformed. This leads to a change in a number of properties: density, thermal conductivity, heat capacity, etc. On this basis, the entire sample can be divided into three zones. In zone 1, the cooling process is in progress, in zone 2 (the reaction zone) there is a combustion process with a large release of thermal energy, which goes to support the combustion reaction and heating zone 3 (preheating zone). In accordance with the methodology used to obtain high-quality samples, the distribution in space and time of new high-temperature combustion centres should be uniform. To verify the fulfilment of this condition by the method of brightness micropyrometry, the dynamics of temperature fields was obtained. The emergence of new foci of combustion of a given temperature was watched by interframe difference. Fast Fourier transform was done in the programs FiJi, OriginPro19. The results are presented in Figures 3, 4.

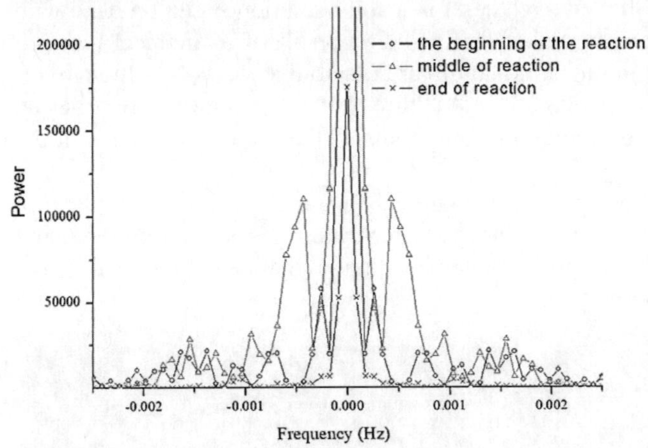

Figure 3. Fourier spectrum of temperature fields corresponding to the beginning, middle, and synthesis process.

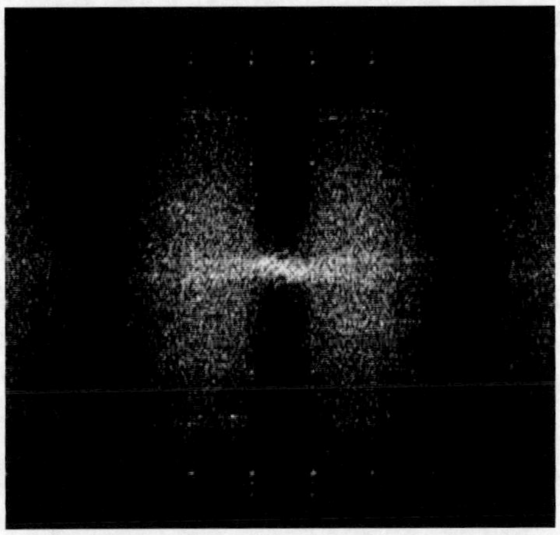

Figure 4. These two figures have been placed side-by-side to save space. Justify the caption.

A comparative analysis of the Fourier spectra of the temperature fields corresponding to the beginning, middle, and completion is observed by a noticeable decrease in temperature at the end of the reaction, and also new highly exothermic regions are reduced. Perhaps for this reason unreacted powder of refractory oxides will be found in the reaction products. Two-dimensional Fourier analysis [7-9] of the entire reaction shows its multistage. So, you can expect in the reaction products the appearance of lower oxides. The lack of highly exothermic foci leads to a slower reaction and its cessation.

Metallographic studies of the initial samples showed a non-uniform distribution of the phase composition, the presence of pores of various sizes. This is a disadvantage in mixing the components of the powder. The reaction of SHS proceeded in a reactor with a diameter close to the critical one. After synthesis, the sample is rapidly cooled due to the large amount of heat sink through the side surface and its small mass (20 g). Melt viscosity increased rapidly with cooling. Gases that did not have time to go out during the phase separation formed large closed pores. The radiograph and microstructure of the sample after the synthesis are presented in Figures 5, 6.

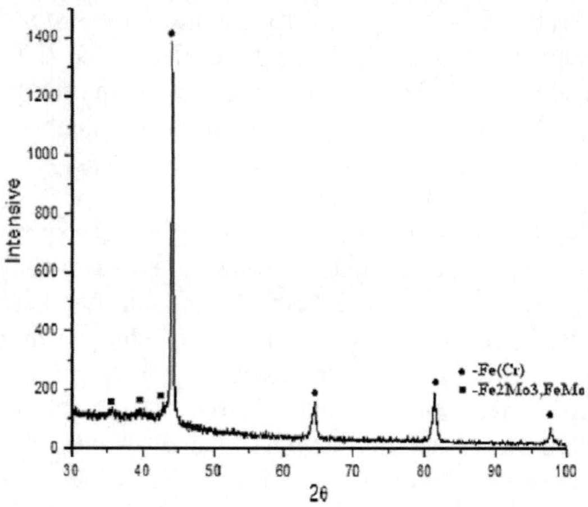

Figure 5. Radiograph of the sample after SH Synthesis.

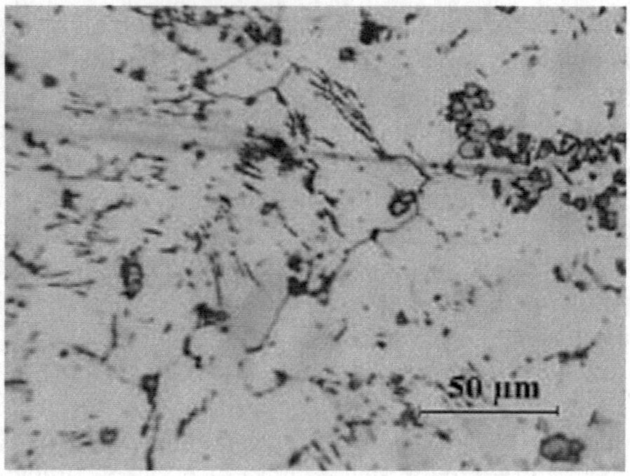

Figure 6. Microstructure of the sample before irradiation.

The structure of the multicomponent system after SH-synthesis is a solid solution based on α-Fe with inclusions of unreacted molybdenum oxide and particles obtained as a result of a chemical reaction between reduced iron and molybdenum. Aluminium and titanium are α-stabilizers, easily soluble in α - Fe and Cr, forming wide areas of solid solutions with a bcc structure. Thus, the solid solution has an increased hardness H_{μ} = 400 MPa due to the dissolution of the atoms of the reduced elements in the iron lattice.

To check the obtained material for radiation resistance, the samples were irradiated with an electron beam. After irradiation with a 1.8 MeV pulsed electron beam, the temperature of the samples remained almost unchanged. The phase composition of the samples after irradiation changed (Figure 7, 8), but the main phase remained a solid solution of α-iron.

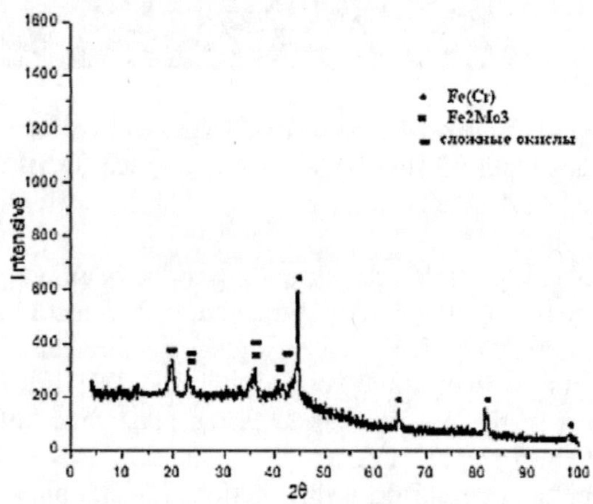

Figure 7. Radiograph of the sample after SH Synthesis.

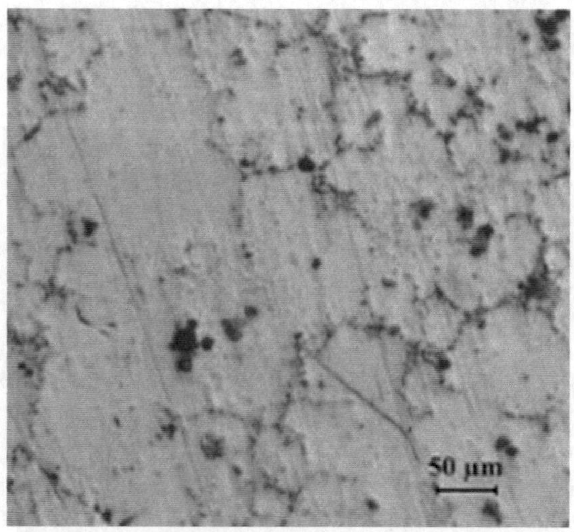

Figure 8. microstructure of the sample after electron irradiation (absorbed dose of 30 Mrad).

With an absorbed dose of up to 20 Mrad, the microhardness of the samples remained almost unchanged.

When an electron beam of more than 1.8 MeV interacts with metals, the main contribution to radiation damage is made by the mechanism of atom ejection and the creation of Frenkel pairs. The calculated migration energy of vacancies in Fe, Cr is from 0.8 to 1.3 eV [10]. The atoms of aluminium and

IOP Conf. Series: Materials Science and Engineering **521** (2019) 012005 doi:10.1088/1757-899X/521/1/012005

molybdenum interact with iron and chromium and form complex oxides. The roentgenogram showed presumably Fe (Cr, Al) $_2O_4$, $FeMo_3O_8$ or FeO * Cr_2O_3. Phase analysis of a sample containing several phases is complicated by the fact that reflections of different phases overlap each other. This means that the same reflex on the diffractogram can belong simultaneously to several phases. At the absorbed dose of 30 Mrad, a small accumulation of pores appears along the grain boundaries. On the surface of the sample, grain sections are well distinguished; their microhardness is H_μ = 421.6 MPa. The crystalline inclusions did not change in size, but the hardness values increased (H_μ = 549MPa). With increasing doses up to 60 Mrad, pores appear in the volume of crystals of the solid solution. previous samples. The crystalline inclusions also increased to 35 μm, and the microhardness values decreased to H_μ = 483.9 MPa. The hardness of the main stages decreased to values of H μ = 404.1 MPa. A sample with an absorbed dose of 70 Mrad was too fragile.
The results discussed may be that the samples withstand electron irradiation. Results can be improved. The distribution of the phase composition in the volume of synthesis products is uneven due to poor charge mixing and incomplete reduction of oxides. Fourier analysis showed that the frequency of occurrence of areas with high heat transfer is insufficient. Although at the beginning of the reaction, refractory elements were recovered from oxides. This is confirmed by x-ray analysis. Thus, an increase in the cost of aluminium should lead to a more complete recovery. Unfortunately, the results of the analysis of the chemical elemental composition are still missing.

5. Conclusion
The alloy was synthesized based on the Fe_2O_3-Al-Cr_2O_3-MoO_3-Ti system. During the research it was found:

1. Fourier analysis of the temperature field of the reaction showed its multistage. the need to increase the mass fraction of thermite mixture.
2. The synthesized material consists mainly of solid solution phases with a bcc lattice based on iron and chromium.
3. Tests of samples for radiation resistance have shown that an increase in the absorbed dose to 30 Mrad leads to the appearance of pores along the grain boundaries.
4. An increase in the absorbed dose to 60 Mrad pores appear in the volume of crystals. In areas rich in chromium, the solid solution disintegrates with the formation of a eutectoid in the chromium-aluminium system.
5. At the absorbed dose, crystalline inclusions increased to 35 μm, and the microhardness values decreased to Hμ = 483.9 MPa. The hardness of the main phase decreased to values of Hμ = 404.1 MPa.

The results indicate the need for further research to improve the method of using Fourier analysis to control phase formation and to establish the optimal ratio of the components of the powder mixture. Each figure should have a brief caption describing it and, if necessary, a key to interpret the various lines and symbols on the figure.

6. References

[1] Rogachev A S and Mukas'yan A S 2015 *Combustion, Explosion, and Shock Waves* **51** 53
[2] Kostin S V, Krishenik P M and Shkadinskii K G 2014 *Combustion, Explosion, and Shock Waves* **50** 42
[3] Garkol' D A, Gulyaev P Y, Evstigneev V V and Mukhachev A B 1994 *Combustion, Explosion, and Shock Waves* **30** 72
[4] Gulyaev P Yu, Cui H Z and Gulyaev I P 2015 *High Temperatures - High Pressures* **44** 83
[5] Song X J, Cui H Z,Cao L L and Gulyaev-P Y 2016 *Transactions of Nonferrous Metals Society of China* **26** 1878
[6] Jordan V, Gulyaev I, Dolmatov A and Gulyaev P 2017 *Journal of Physics*: Conference Series **830** 012151
[7] Dolmatov A V, Gulyaev I P. Gulyaev P Yu and Iordan V I 2016 *IOP Conference Series: Materials Science and Engineering* **110** 012058
[8] Gulyaev P., Jordan V. Y 2017 *Workshop Proceedings* **37** 1940

IOP Conf. Series: Materials Science and Engineering **521** (2019) 012005 doi:10.1088/1757-899X/521/1/012005

[9] Song X J, Cui H Z, Cao L L and Gulyaev· P Y 2016 *Transactions of Nonferrous Metals Society of* China **26** 1878
[10] Lua X, Naidisb G V, Laroussic M and Ostrikov K 2014 *Physics Reports* ***540*** *123*

Acknowledgments

This work was supported by the Russian Foundation for Basic Research in the framework of scientific projects No. 18-08-01475, No. 18-41-22004 and project No. 13-01-20 / 28 of the leading scientific school of the Yugra State University.

Research on Construction Process of the Cable-net Structure of FAST

Xu Kong [1,2,3,a], Qiming Wang [4]

1 China Manned Space Agency, Beijing 100034, China
2 Manned Space Technology System Center, Beijing 100094, China
3 Tsinghua University Department of Industrial Engineering, Beijing 100084, China
4 National Astronomical Observatories Chinese Academy of Sciences, Beijing 100012, China

[a] Corresponding author: kongxu@bao.ac.cn/kongxu@mail.tsinghua.edu.cn

Abstract. With a huge collecting area of more than 30 football fields, Five-hundred-meter Aperture Spherical radio Telescope (FAST) is the largest single dish in the world. There are many construction methods of large span cable-net structures some of what is very complicated, so the distribution of internal force of structures is intricate and alternative. According to a variety of mechanical problems and technical difficulties existing in the construction process determine the general construction scheme of the podium, through simulated the construction process, optimize the construction program. The research showed that it had no effect on the structure to consolidate part of the construction steps of theory construction scheme, the actual construction scheme to be used is safe and feasible, and the important conclusion that cable crane and V-zone system are safe is arrived.

1. Introduction

Five-hundred-meter Aperture Spherical radio Telescope (FAST) is supported by cable-net structure, which enables its surface to form a paraboloid in real time under active control. It belongs to the national "11th Five-Year Plan" projects, the project time is 5.5 years from the commencement of work in March of 2011 to 2016. [1, 2]

The structure of the cable-net of FAST is a spherical dish with a radius of 300m and an opening up to 500m in diameter. It is located at karst geomorphology in China, so the construction area is limited and the large type hoisting machinery is difficultly used. The cable-net of the active reflector of FAST is a complex structure system composed of about 6670 main cables segments, 2225 down-tie cables, 50 latticed pillars and steel ring beam of diameter 500m, as shown in Fig.1. In order to simplify the structure of the main reflector, by reducing the number of movable parts and avoiding heavy civil engineering between actuators and the ground. The effective aperture of 300 m will be illuminated by the feeds in a cabin moving on the focus surface, halfway from the reflector to its spherical center. The telescope is to be pointed by moving the focus cabin and adjusting simultaneously the shape of the illuminated reflector from a sphere to a paraboloid. [3, 4, 5]

There are many construction methods of large span structures some of what is very complicated, so the distribution of internal force of structures is intricate and alternative. We should determine the reasonable order during the process of construction to make sure the cable-net construction go safe and economical. It is necessary to simulate the whole construction process of large span cable structures to make out the mechanics property of structures and study the relevant problems during this process. [6,7]

Figure 1. FAST engineering

In this paper, systematically and thoroughly study and evaluations were carried on the construction mechanics of the cable crane, basing on the cable-net structure of FAST. The research methods include numerical simulation, similarity model test and structure analysis based on key area. Based on structure cases, this paper recommends rationally arranging construction procedures and fully utilizing structural characteristics， the analysis shows that if the above mentioned methods are accepted, and the construction is processed according to the optimized process, a shortened construction period, as well as a lowered construction cost, will be obtained. At present, firstly the independent towers were installed in the middle of cable-net in order to assemble of main cable and down-tie cable. Secondly, the installation of cable began form the cable of five-symmetry axis-plan by virtue of catwalks and cable crane. Finally, the height control is the key points to the construction quality and to ensure the success of cable-net closure.

2. The optimization of construction method

FAST project is currently in the construction and implementation stage, the construction method is one of the feasible solutions for cable-net structure, which is an effective supplementary type for the same structure by using cable crane and catwalks, some outstanding problems caused by large span space structures such as complicated topography, large land area, high slope, and so on, can be solved more effectively. However, the application and study of the cable-net are still at a preliminary stage. The systematic study on the mechanical response during construction is fairly few. In structure practice, the guideline for design and construction is insufficient. [8]

The cable-net of the active reflector of FAST is a complex structure system composed of about 6670 main cables segments, 2225 down-tie cables, the main cables are almost 12m long and the driving cables are 3m to 50m long， especially in the middle area the most cable of length are less than 5m. The geodesic meshed cable-net structure has been designed as the supporting structure of FAST, and five-symmetry axis-plan was thus suggested as preferred mesh plan. According to the distribution of the cable-net, main cables and down tied cables are divided into five areas. To diminish the harm caused by the high altitude working platform, consider in the middle area the most cable of length are less than in construction procedure we have made a comparison with standing alone tower directly built on the ground. From this respect, according to the region terrain, the geology and the environment situation of constructs section, the cable-net of FAST can be divided into construction area， in which various support plans such as catwalks, cable crane and hanging basket and so on. The construction sequence of cable-net structure the main reflector of FAST, to be shown up that from left to right panel of Figure 2.

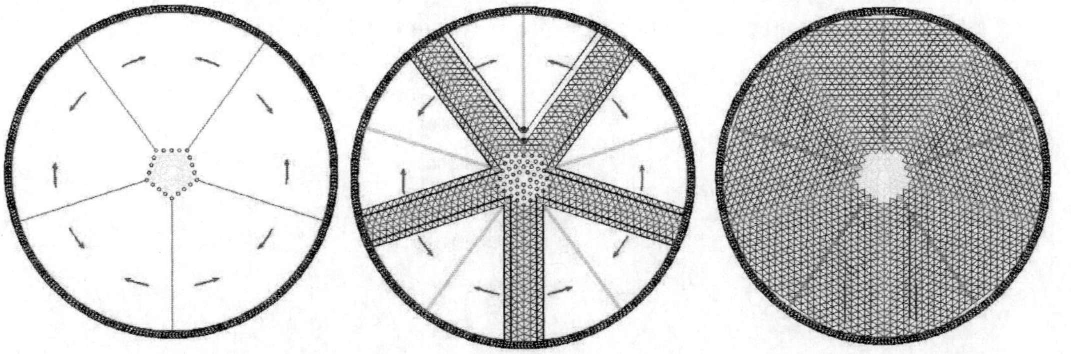

(a) (b) (c)

Figure 2. The construction sequence of the cable-net of FAST

2.1. Simulating analysis on the mechanics theory during the process of construction

This paper point out the control factors and establish the steps of the mechanics analysis of the cable-net structures during construction process. The static force analysis during simulating construction process of the cable-net of FAST, we establish the three dimensions model by means of finite element analysis, simulate the statics property of steel roof during construction process by the effect of dead load, live load and disadvantage accidental action. Then select the reasonable construction scheme and make the relatively economical result and satisfy the structure design scheme at the same time. [9]

The non-linear analysis of space cable structures by the imperfection of construction, the structures will generate geometric or material imperfection in some degree, embody mainly in the variety of distribution of internal force in structural units and the geometric imperfection of key nodes of structures. Compare the result of construction analysis with testing and ideal design model, assess the effect of practical construction process on the formation of cable-net structures, then offer some suggest on the readjust of design and reasonable construction scheme.

2.2. Modelling cables and cable networks

Different numerical approaches have been proposed to model cable structures. In particular, there are various catenary-type finite elements as well as analytical elements available, all stemming from the direct problem-solving of the equilibrium and compatibility equations.

The cable elements adopted are infinitely stiff in the axial direction and cannot experience any increment of length. In fact, the catenary is defined as the curve which an inextensible cable, with negligible resistance to bending, adopts when suspended from its ends and is subject to the effects of gravity. Its mathematical expression has been known for a long time. In 1691, in response to a challenge set forward by Jacob Bernoulli, the scientists Leibnitz, Huygens and Johann Bernoulli found the equilibrium equation for an inextensible catenary. Leibnitz used the theory of infinitesimal calculation to derive the equation of this curve.

In practice, computer applications based on this type of element encounter severe difficulties due to the fact that the curve that relates forces and displacements, in the inextensible catenary, is asymptotic. Solving procedures tend to experience large numerical instability causing, in turn, a very difficult or even impossible convergence.

FAST cable-net is formed using geodesic method, the cable-net is of 1/5 symmetry about the vertical line through the center of the sphere cap. The main cable at symmetrical positions can be classified into a same group of same specifications. Considering the 1/5 symmetry of the cable-net, the calculation load can be effectively reduced. For the cable crane, there will be a calculation case and the cable stresses corresponding to each cable segment are stored, there are 22 working conditions in sum. Based on statistical data of useful life of tractive-rope and guide-rope gathered from 4 material simulate study as well as practical experiences in the field, problems, such as rope and ropeway profile selection, running conditions and maintenance works, are discussed with recommendations for rational use of tractive-rope for designer, operators and maintenance workers reference. For the calculation case, the force of the tractive-rope and guide-rope are shown in Fig.4(a) and (b).

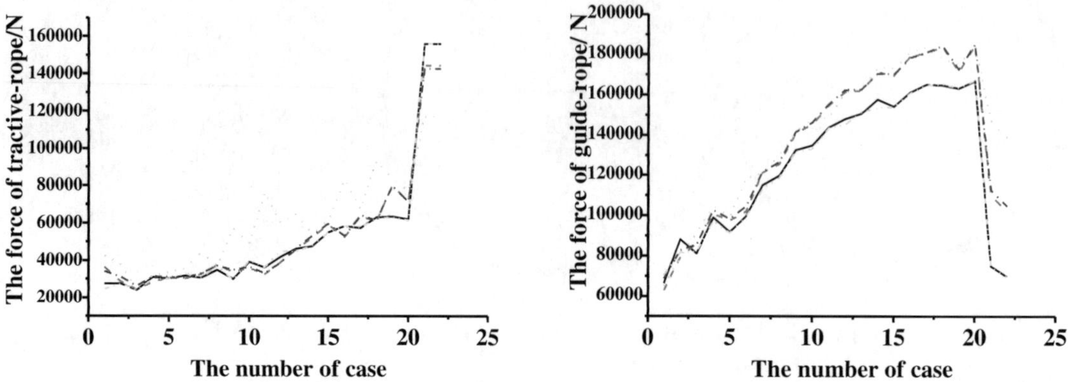

(a) Tractive-rope (b) Guide-rope

Figure 3. Simulation force results of tractive-rope and guide-rope

2.3. The static and stable analysis of V-zone system

According to the construction control, combining to the catwalk of cable-net, the V-zone was be designed and used to control the anchoring point in complicated system.It is necessary to analysis the statics and stability property of large span cable structures because it's importance and the complexity of it is force transmission path system. Ascertain the V-zone system by compare some candidate construction scheme. This paper study on the static and stable property of V-zone system then put forward a kind of assembly all-purpose V-zone systems. The optimize principle of this method is that the internal force in structure and V-zone system is minimal and the process of discharge shortest. Finite Element Model of V-zone is shown in Figure 4 a (left panel), and stress variation distribution of V-zone structure is shown in Figure 4 b (right panel).

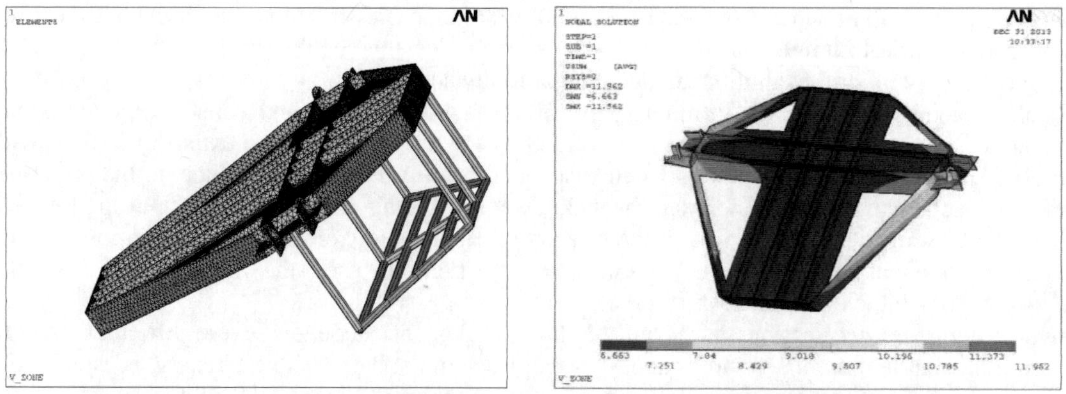

(a)Finite Element Model of V-zone (b) Stress variation distribution of V-zone

Figure 4.V-zone system structure

3.Conclusion

This paper takes the construction mechanical analysis as means, the cable-net structure of FAST as the research object, the key technical problems during the construction process of the large span cable-net structure have been studied, and the following conclusions are obtained:

(1).Construction method of cable-net structure supporting the reflector of FAST is discussed, based on optimization principle of construction methods, and also excavation effects induced by construction sequences and numerical simulation are analyzed by FEM, adopting the typical method of large span

cable structures. It is indicated that existing construction methods are reason able during construction, the basis and guidance are supplied for real engineering and some meaningful conclusions are gotten as the reference for similar engineering.
(2).Put forward some key technologies during the installation process of the podium, and make some analysis and studies, the analyzes deformation of the V-zone structure during cable-net installing procedure and study the impact of deformation of precise location when splicing the cable-net at high altitude and determine the construction scheme.

4. References

[1] Nan Rendong, Ren Gexue, Zhu Wenbai, 2003, Acta AstronomicaSinica,44,13-18
[2] Nan Rendong, Di Li, Chengjin Jin, 2011, International Journal of Modern Physics D,20, 6
[3] Nan Rendong, 2006, Science in China: Series G, Physics, Mechanics and Astronomy, 49,2
[4] Li G Q, Shen L, Luo Y F, 2011, Ap&SS, 278,1
[5] Jiang P, Wang Q M, Zhao Q, 2011, Applied Mechanics and Materials, 94-96,979~982
[6] QIAN H L, Fan F, SHEN S Z, and WANG Q M, 2005, Journal of Harbin Institute of Technology, 37,750~752
[7] Qian Hongliang, 2007 ,Harbin Institute of Technology(2007)
[8] Kong X, Jiang P, Wang Q M, 2013, Engineering Mechanics, S1,169-174
[9] Banan M R, Hjelmstad K D, 1994, Parameter Estimation of Structures from Static Response. Ⅰ: Computational Aspects,Journal of Structural Engineering，ASCE, 120,11

Acknowledgments

This work was supported by the National Natural Science Foundation of China (Grant No.11803033). We would like to thank all our colleagues for their contributions to the study of cable-net structure for FAST.

IOP Conf. Series: Materials Science and Engineering **521** (2019) 012007 doi:10.1088/1757-899X/521/1/012007

Machining of Carbon Fibre Reinforced Polymer Composite by Electrochemical Discharge Machining Process

Manpreet Singh[1#], Sarbjit Singh[2]

[1, 2] Punjab Engineering College (Deemed University), Chandigarh-160012, India

#Corresponding author's e-mail: gurayamanpreet5800@gmail.com

Abstract. Carbon Fibre Reinforced Polymer (CFRP) composite is highly applicable in aeronautics, defence and electronics industries application due to their better mechanical properties. Still, the micro-drilling of CFRP composite is challenge to research society due to its semi-conductive and fibrous nature. In this work, the author's successfully drilled micro-holes in CFRP composite using Electrochemical Discharge Machining (ECDM) process. The experiments were designed on the basis of Taguchi's method using L_9 orthogonal array. The experiments were carried out with process parameters such as electrolyte concentration, applied voltage, inter electrode gap and material removal rate as response characteristic. The surface characteristics of drilled hole were studied by Scanning Electron Microscope (SEM) micrograph. Additionally, the dominant effects of process parameters on MRR were evaluated using Analysis of Variance (ANOVA).

1. Introduction

In recent years, the Polymer Matrix Composites (PMC's) have attained popularity in aeronautics, defence and other applications due to low weight, high strength and less thermal expansion [1]. The machining of PMC's is challenge to research society due to different irregularities occurred while machining by traditional machining methods. With development of advance materials, the researchers tried to develop a process which can machine various materials without constraint to their properties. The Electrochemical Discharge Machining (ECDM) process has gained reputation in the machining of conductive and non-conductive materials irrespective to their properties [2]. ECDM process is firstly attempted to machining of glass by Karafuji [3]. Later on, ECDM process is successfully attempted to machining of composites [4], ceramics [5], steel plates [6] and super-alloys [7]. ECDM process is hybrid machining method which pools the process individualities of Electrochemical Machining (ECM) and Electrical Discharge Machining (EDM) [8]. The ECDM setup includes two electrode terminals i.e. tool electrode and auxiliary electrode alluded to as cathode and anode independently as showed up in **Figure 1**.During machining, the cathode is submerged up to couples of millimetres in electrolyte solution and anode electrode is arranged at few centimetres away from tool electrode tip. The material to be machined is positioned at vicinity of cathode and submerged in electrolyte. With initiation of DC supply, the hydrogen bubbles starts formed on tool electrode surface due to electrolysis. The high current density at cathode tip is stepped up generation rate of hydrogen bubbles and combines to form a gas film layer. The layer casing formed on tool electrode surface behaves as a dielectric and eliminates contact of electrolyte with cathode. When the applied electrode potential increases beyond the disruption strength of gas film, the discharge is produced at periphery and tip of cathode electrode. The machining material in the vicinity of cathode electrode gets softening, vaporization and thermal spalling caused by spark energy [4].

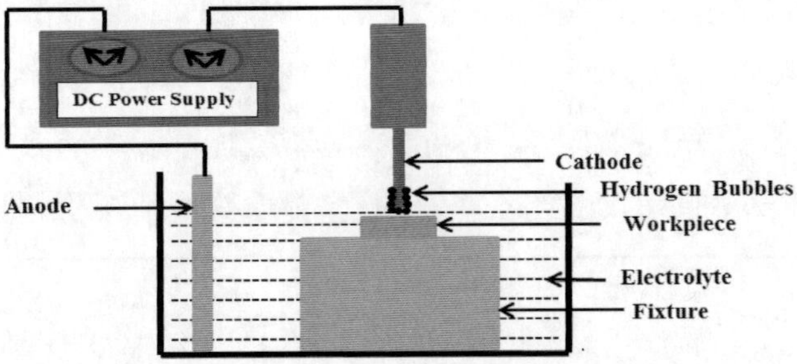

Figure1: Schematic of ECDM process

In domain of composites, Liu et al. [9] conducted micro-machining of conductive composite material using grinding aided ECDM. They concluded that nature of material removal criterion in ECDM is equivalent to EDM i.e. spalling of debris. Manna and Narang [10] attempted micro-drilling of E-glass fibre epoxy composite by ECDM process. They concluded that, the machined surface characteristics were non-uniform due to imperfect machining of fibres. Antil et al. [11] effectively machined micro-holes in SiC reinforced PMC's. They observed heat affected zone (HAZ) and micro cracks on machined surface by SEM micrographs. From the available literature, it is revealed that machining of various materials such as composites, ceramics, steel and super-alloys have been reported by different researchers. However, the machining of carbon fibre reinforced polymer composite (CFRP) by ECDM process is least reported. The principle point of this study is to examine the impact of process parameters such as applied voltage, electrolyte concentration and inter-electrode gap on response characteristic i.e. material removal rate (MRR). Additionally, to find out optimum parametric conditions for machining of CFRP composite using ECDM process.

2. Experimentation and Methodology

The material used for experimentation purpose is carbon fibre reinforced polymer composite (CFRP) is procured from Indigenous Ltd., Dehradun, India. The characteristics of material such as warp fibre-carbon fibre 3K 200tex, weft fibre-carbon fibre 3K 200tex. The weight distribution of carbon fibres in CFRP composite is 50% warp: 50% weft with an orientation of 0/900. The thickness of work piece material is 2mm. The micro-drilling of CFRP composite is carried out on developed ECDM set up. The machining parameters for experimentation were applied voltage (50-70V), electrolyte concentration (30-50%) and inter-electrode gap (60-80mm) with performance characteristic such as material removal rate (mg/min).The input parameters values for experimentation with their levels are shown in **Table 1**. The NaOH electrolyte is chosen among other electrolytes due to high popularity in ECDM process [8]. While machining, each experiment is carried out by first hand carbide tool drill bit of diameter 300μm. The experiments were performed by using Taguchi's L9 Orthogonal array. The output response characteristics of nine alternative experiments are shown in **Table 2**. The material removal rate (MRR) in each experiment is calculated by weighing machining sample with reference to time.

Table 1: Process parameters and their Levels

Symbols	Process Parameters	Level 1	Level 2	Level 3
A	Electrolyte Concentration (Mass Percent)	30	40	50
B	Applied Voltage (Volts)	50	60	70
C	Inter-Electrode Gap (mm)	60	70	80

Table 2: Experimental results by Taguchi's L_9 Orthogonal array

Sr.	Electrolyte	Applied Voltage	Inter Electrode Gap	MRR (mg/min)

No.	Concentration			Mean	S/N Ratio
1	30	50	60	1.102	0.84363
2	30	60	70	1.192	1.52553
3	30	70	80	1.210	1.65571
4	40	50	70	1.178	1.42291
5	40	60	80	1.195	1.54736
6	40	70	60	1.280	2.14420
7	50	50	80	1.165	1.32652
8	50	60	60	1.295	2.24540
9	50	70	70	1.310	2.34543

3. Results and Discussion

The experiments were performed on developed ECDM set up with L_9 Taguchi's orthogonal array. The signals to noise ratio and raw data plots for MRR are shown in **Figure 2**. The graphs illustrates that MRR is increased linearly from level A1 to A3 of electrolyte concentration. The high concentration of electrolyte is steeped up the electrochemical reactions in between cathode electrode and anode electrode of ECDM process [4]. Therefore, the large amounts of hydrogen bubbles were formed in machining zone due to accelerated electrochemical reactions. This phenomenon increases the rate of sparking in machining zone. This finding is justified from previous research work in machining of glass fibre reinforced PMC's using ECDM process [4]. Similarly, the MRR is also increased with rise in applied voltage in ECDM process. On the contrary, the MRR is firstly increases and then shows inverse trends with increase in inter-electrode gap of ECDM process. The higher inter-electrode gap increases the inter-electrode resistance while machining which step down the intensity of sparking in machining zone [11]. The main focus of this study is to find out optimum parametric values for MRR while machining CFRP composite. From these results, the optimal values for MRR are A3B3C2 i.e. 50 (Mass Percent), 70 V and 70 mm. Additionally, the analysis of variance (ANOVA) for MRR are shown in **Table 3**.ANOVA depicts that electrolyte concentration and voltage remains an dominant factors for MRR in machining of CFRP composite using ECDM process. The percentage contribution of process parameters on MRR are graphically represented in **Figure 3**.

Table 3: Analysis of variance for MRR

Source	Symbols	DF	SS	MS	P Value	Percentage contribution
Electrolyte Concentration	A	2	0.011850	0.005925	0.076	31.82
Voltage	B	2	0.021791	0.010895	0.043	58.52
Inter-electrode Gap	C	2	0.002618	0.001309	0.272	7.03
Error		2	0.000977	0.000488		2.62
Total		8	0.037235			

The SEM micrograph of machined hole on CFRP material at 150x magnification is shown in **Figure 4**. The SEM micrograph depicts drilling of through hole with complete cutting of fibres from surface of composite. Moreover, the periphery of hole is slightly non-uniform due to semi-conductive nature of CFRP composite. Apart from these, cracks and uneven surface produced on outer surface of drilled hole. This is because of limited supply of fresh electrolyte while machining. The supply of fresh electrolyte in machining regime produced continuous and sparking phenomenon. In future, the non-uniformity of machined surface can be addressed by incorporation of other techniques such as vibration assistance, tool electrode geometry etc. which promotes flushing of electrolyte in machining zone.

IOP Conf. Series: Materials Science and Engineering **521** (2019) 012007 doi:10.1088/1757-899X/521/1/012007

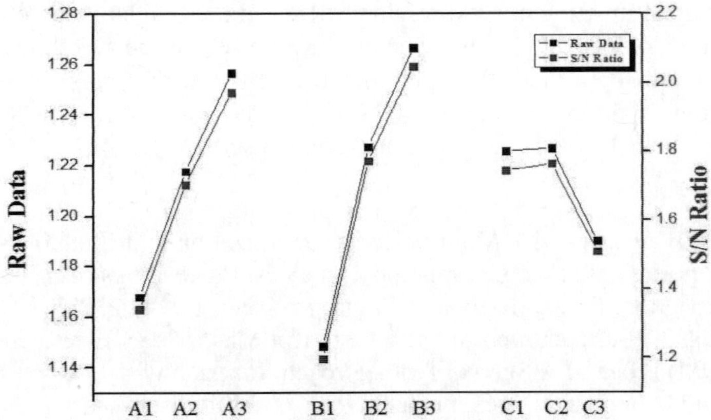

Figure 2: Raw data and S/N ratio for MRR

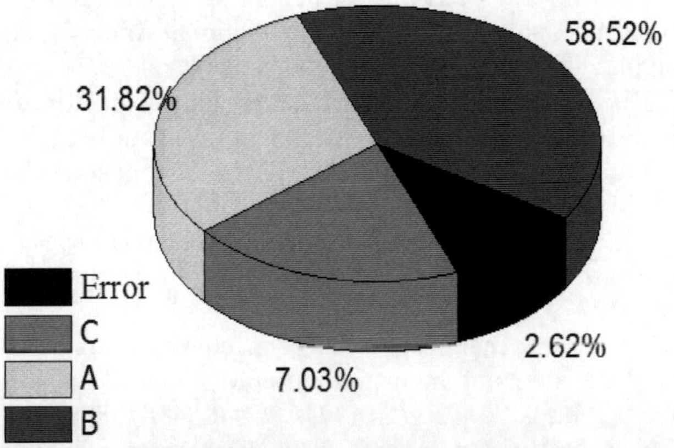

Figure 3: Contribution of process parameters for MRR

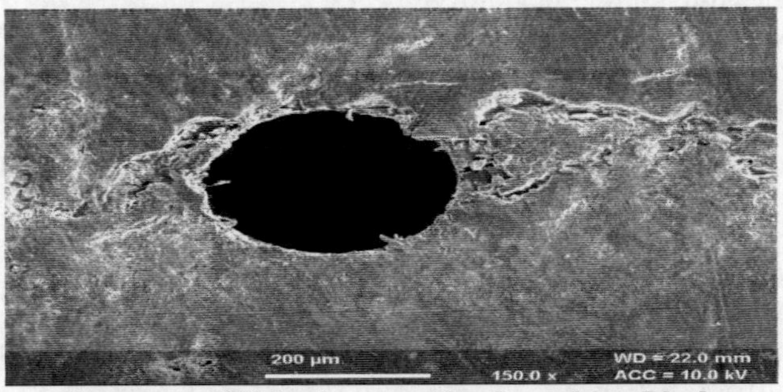

Figure 4: SEM micrograph of machined hole

4. Conclusions

Based on experimental results in machining of CFRP composite by ECDM, the following conclusions were drawn:

- The ECDM process is effectively used for machining of polymer matrix composites irrespective to their properties.
- Based on Taguchi's technique, the optimum parametric values for MRR are electrolyte concentration, applied voltage and inter-electrode gap are 50% (mass percent), 70V and 70mm respectively.

- The experimental investigation shows that MRR is directly related to voltage and electrolyte concentration of ECDM process, whereas the inter-electrode gap shows inverse trends.
- SEM micrograph depicts drilling of hole on CFRP with fibres removal on surface of composite by sparking phenomenon. However, the outer surface of drilled hole becomes less uniform due to cracks and uneven machined surface.

5. References

[1] Singh S, Singh I, Dvivedi A 2013, Multi objective optimization in drilling of Al6063/10% SiC metal matrix composite based on grey relational analysis, Proceedings of Institutions of Mechanical Engineering, Part B: Journal of Engineering Manufacture, 227(12),1767-1776.

[2] Singh M., Singh S. 2018, Machining of Difficult to Cut Materials by Electrochemical Discharge Machining (ECDM) Process: A State of Art Approach. In: Antony K., Davim J. (eds) Advanced Manufacturing and Materials Science. Lecture Notes on Multidisciplinary Industrial Engineering. Springer, Cham

[3] Karafuji H, Suda K, 1968, Electrical discharge drilling of glass, CIRP Ann. ,16, 416-419.

[4] Antil P, Antil P, Singh S, Manna A 2018, Electrochemical Discharge Drilling of SiC Reinforced Polymer Matrix Composite Using Taguchi's Grey Relational Analysis. Arabian Journal for Science and Engineering.,43(3):1257-66.

[5] Sarkar BR, Doloi B, Bhattacharya B 2006, Parametric analysis on electrochemical discharge machining of silicon nitride ceramics, Int. J. Adv. Manuf. Technol., 28, 873-881.

[6] Coteata M, Slatineanu L, Dodun O, Ciofu C 2008, Electrochemical discharge machining of small diameter holes, Int. J. Mater. Form , 1, 1327-1330.

[7] Zhang Y, Xu Z, Zhu Y, Zhu D 2016, Effect of tube-electrode inner structure on machining performance in tube-electrode high-speed electrochemical discharge drilling, Journal of Materials Processing Technology ,231, 38-49.

[8] Singh M, Singh S. 2018, Electrochemical discharge machining: A review on preceding and perspective research. Proceedings of the Institution of Mechanical Engineers, Part B: Journal of Engineering Manufacture, https://doi.org/10.1177/0954405418798865.

[9] Liu JW, Yue TM, Guo ZN 2010, An analysis of the discharge mechanism in electrochemical discharge machining of particulate reinforced metal matrix composites. International Journal of Machine Tools and Manufacture, 50(1):86-96.

[10] Manna A, Narang V 2012, A study on micro machining of e-glass–fibre–epoxy composite by ECSM process. The International Journal of Advanced Manufacturing Technology, 61(9-12):1191-1197.

[11] Antil P, Singh S, Singh PJ 2018, Taguchi's Methodology Based Electrochemical Discharge Machining of Polymer Matrix Composites. Procedia Manufacturing.26:469-73.

Acknowledgments

The author's acknowledge the financial support provided by Science and Engineering Research Board under the Department of Science and Technology, New Delhi, Government of India, for this research work under grant EMR/2016/005352.

IOP Conf. Series: Materials Science and Engineering **521** (2019) 012008 doi:10.1088/1757-899X/521/1/012008

Alarm Audit and Enforcement: Automating the Integrity of Alarms for Manufacturing Systems

T. Ault

Automation Engineer & Alarm Advisor, Chevron & Colorado State University, 709 Lenox Ct. Paso Robles, CA, 93446, USA

L. Yang

Dept. of Electrical & Computer Engineering Colorado State University, 1373 Campus Delivery Fort Collins, CO 80523-1373

Abstract. This work details the development of a continuous monitoring system technology to track the compliance of alarms in a manufacturing environment. A problem arose due to having simultaneous capital projects and continual process optimization work at several production sites. With continual changes to the control systems, a project was developed to ensure that alarm adjustments never violated established safety thresholds for systems. A technique was used to rationalize alarms using a risk based objective analysis to establish safety thresholds for each alarm. Then a tool was developed that continually cross-referenced alarm safety server with process control network to immediately notify management and engineering of any adjustments that violated alarm and safety rules for the site. This work led both to improved site safety and to an improvement of alarm metrics due to greater confidence in adjusting alarms with this audit tool. One site saw a monthly reduction of the quantity alarms by 55% and the elimination of alarm floods.

1. Introduction

Several production sites in southern California were undergoing major capital upgrades, large amounts of maintenance work, and optimization projects concurrently. At any given time, there would be several programming contractors either installing a new vessel with associated controls, updating process control network (PCN) databases, or tuning control valves in an industrial setting. In the past, alarms were hardwired and required significant planning to remove or place new alarms. With the widespread use of distributed controls systems (DCS), the amount of alarms has increased at sites along with the ease to alter them [1]. Due to simultaneous operations with construction, operations, and maintenance, it was a required that an additional safety system was needed to alert engineering and management if any new alarms were added, deleted, or modified that would endanger the sites. It was determined to continually cross reference an alarm master database with process controllers to determine instantaneously if any unsafe conditions exist with regards to alarms. Alarm auditing the process control network (PCN) against an alarm master database has been utilized in industry and the lack of such a system has been attributed as a contributing factor to some process incidents [2]. Few examples were found for the use of a continual monitoring system for the alarms. The continual monitoring poised some technical and organization challenges in this project.

The general established method for alarm rationalization is based on American National Standards Institute (ANSI) and International society of automation (ISA) [3]. There are several modifications to this technique for alarm rationalization ranging from the very simple to the complex [4]. Alarm

IOP Conf. Series: Materials Science and Engineering **521** (2019) 012008 doi:10.1088/1757-899X/521/1/012008

rationalization is a systematic approach to evaluate every alarm tag in a facility to determine whether it is an alarm and if so document consequences, impact, corrective actions, response time, and priority. Inconsistent and undisciplined use of alarms can create a frustrating situation for operations and lead to ignored alarms. The goal of alarm rationalization is to review, validate, and justify alarms that meet the criteria of an alarm. The goals include the most efficient number of alarms to ensure the process system is safe, operators have sufficient time to respond, and operates in a safe process range. [5] One of most important parts of alarm management is considered alarm rationalization and the approach used will be a major element if the effort is successful or fails to manage alarms at a large site. [6] A risk-based approach was utilized to evaluate each proposed alarm and risks associated with it to determine key parameters of each alarm. The parameters to determine during the analysis included maximum time to respond (MTR), alarm priority, alarm type, cause, operator actions, and setpoint. After the alarm rationalization was completed, a master alarm database was created with the results. From there an automated alarm audit and enforcement tool was created to cross reference the actual process control network alarm settings with the alarm rationalization in the master alarm database. While the continuous audit of online systems may seem easier due to process control and network surveillance being more common and continually improving, challenges always exists getting databases and software to communicate with each other. Varsahelyi et al discuss some of the historical evolution of continually monitoring large amounts of online data. [7] This paper reviews the unique approach used for alarm rationalization and the development of the automated alarm audit and enforcement tool. Results document the improvements in alarm metrics and related work that occurred because of the project.

2. Methodology

Management determined that the current state of alarm management should be improved to prevent the unauthorized changes of alarms which could violate the safety thresholds of the sites. This was a concern due to simultaneous programming projects occurring at facilities undergoing capital improvements by several programming contractors. The first step was to rationalize the alarm to determine the correct parameters. While data existed for some sites for the alarm tag, setting, priority, and other important data, it was requested to refresh the data at each site. It was agreed that for all new systems, modified systems, and systems with known problems a formal process will be undertaken to document and correct alarm rationalization issues. A systemic approach was developed to rationalize the alarms based on an objective analysis carried out by facilitator and key personnel. This approach was termed an alarm objective analysis (AOA). The following participants were considered key participants for the objective analysis: Facilitator, head operator, shift operator, process engineer, programmer, automation engineer. Engineering documentation is also required for the meeting which included (but not limited to) piping & instrumentation diagrams (P&IDs), process hazard analysis studies (PHA), cause and effect matrix, and list of alarm tags. With the key personnel and engineering data collected, the goals of the alarm analysis are:

- Identify, rationalize, and document alarms and alarm setpoints that are required for safe operations
- Select minimum number and proper type of alarms
- Define unique responses to alarms
- Assign alarm priorities based on the severity of the potential abnormal condition and the maximum time to respond (MTR)
- Develop database the defines the required alarms

To aid this, a spreadsheet was developed for each alarm where the initiating event could be recorded with the alarm to respond to the event. From there the maximum time to respond could be analysed to determine the priority of the alarm. The spreadsheet recorded alarm tags, causes to trigger alarm, maximum time to respond (MTR), setpoint, engineering units, and severity of cause. With this data captured, the team can utilize the simplified table shown (Table 1) to determine the proper alarm priority.

Table 1: Simplified table for alarm consequences

	Consequence			
MTR (minutes)	None	Minor	Major	Severe
>30	No Alarm	No Alarm	No Alarm	No Alarm
5-30	No Alarm	Low	Low	High
1-5	No Alarm	Low	High	High
<1	No Alarm	High	Urgent	Urgent

It should be pointed out that "no alarm" for this example means that the operator is still notified but it does not get elevated to an audible and high alert visual notification. Typically, an incident that has a high maximum time to respond and low consequence was re-engineered. This is due to if the incident is a slowly developing process, safety controls could be introduced to control the risk. Another example of alarm prioritization matrix is shown in table 2. These tables and risk tolerances all vary between industries, it is important only to have an agreed set of alarm priorities that are aligned with the risk to the site.

Table 2: More detailed table of alarm consequences

Alarm Priority Determination						
	Severity of Consequences					
Time Available	Incidental	Minor	Moderate	Major	Severe	Catastrophic
> 60 Min	NO ALARM	NO ALARM	NO ALARM	Re-Engineer	Re-Engineer	Re-Engineer
30 – 60 Min	NO ALARM	LOW	LOW	LOW	HIGH	HIGH
10 - 30 Min	NO ALARM	LOW	LOW	HIGH	HIGH	CRITICAL
< 10 Min	NO ALARM	LOW	HIGH	HIGH	CRITICAL	CRITICAL

Based on the results of the alarm objective analysis, the following set of process control network priorities were also developed to further assist operations organize and manage alarms (table 3). These are the priorities that will be used on the human machine interfaces (HMIs) to alert operators and list the current active alarms. The lowest number (more critical alarms) will be listed above other alarms that are current occurring. Table 3 below lists the relationship between alarm priorities from the alarm objective analysis and what was used in programming to further organize the alarms.

Table 3: Developed alarm priorities based on alarm consequences

Alarm Priority System	
Alarm Priority	Controller Alarm Priority
CATASTROPIC	1
CRITICAL	2 to 99
HIGH	100 to 199
LOW	200 to 499
JOURNAL ENTRY	500 to 799
DIAGNOSTIC	800 to 998
NO ALARM	999

The distribution of alarm priorities should be reviewed periodically to ensure that the risk profile of a site is not skewed to highly to urgent alarms or low priority alarms. To provide a suitable risk profile for the plant there should be a reasonable amount of urgent alarms to manage. In other words, too many high urgency alarms will lead to distraction and not capture the true risks to the site. Like other safety analyses, the distribution of alarms from catastrophic, to critical, to high, and then to low should resemble a pyramid. An example of this is given below in table 4 which was developed from one of the upgraded sites. The middle column shows the current percent of each alarm priority along with what the project team thought the recommended alarm priority quantity should be for the sites.

Table 4: An example of the profile of alarm priorities for a sample site

Alarm Priority	Current	Recommended
Most Urgent	4%	2-10%
Medium	16%	10-25%
Least Urgent	80%	65-80%
Total	3201	

The below figure (figure 1) summarizes the steps of the alarm objective analysis (AOA). Due to either a new project being commissioned, lapse in time since last analysis, or other reason, an AOA would be conducted, and a spreadsheet developed (Step 1). The engineering data would be fed into the data and validated (Step 2). The meeting would occur and based on alarms and their event consequences priorities would be established (Step 3). The results would be reviewed and approved (Step 4). Finally, the documentation would be collected and uploaded into a master alarm database (Step 5). After this was completed each site had an approved alarm master database that included the critical data needed for the alarm audit and enforcement tool.

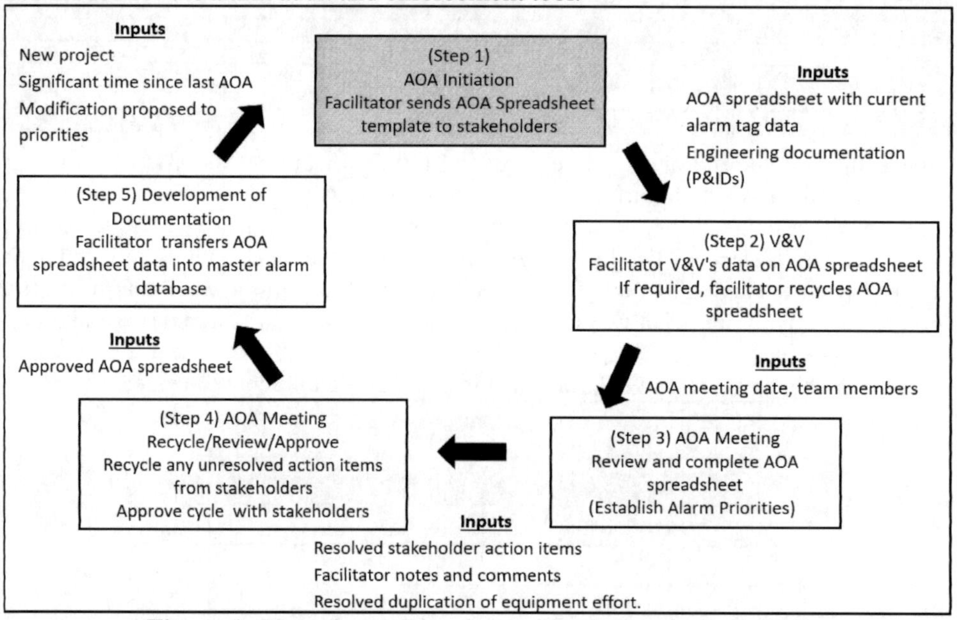

Figure 1: Flow chart of the alarm objective analysis (AOA)

The next stage of the project was to design the architecture to communicate between the process control network and the alarm master database (also termed the alarm safety server). The goal of the development was to provide an automated reconciliation audit and enforcement process to verify that alarm settings implemented in site's process control network (programmable logic controllers, human machine interfaces) matched the designed settings stored in the master alarm database. The process will run on a continual basis to identify changes made to alarm set points and priorities. It was also designed to determine if alarms were deleted or added at the process control network without update the master alarm database. If automated enforcement is enabled, discrepancies shall be rectified by changing the settings implemented in the process control network to match the master database settings. If automated enforcement is not enabled, discrepancies shall be rectified by changing the implemented settings to match the master database settings manually. System change audit logs will be created by the reconciliation process to document the discrepancies found. The development to enable the communication between the HMIs, historian, and alarm safety server began with creation of alarm audit tag objects are created by the custom master alarm audit object during initial configuration, and become a static resource once deployed. Plant expansions may require that more objects be created; in that case the master alarm audit object must be reconfigured for the larger number of alarms. A custom master alarm audit object was created using Visual C# and tool kits provided by software used in process control network. These objects scan the system for alarm objects, and create, configure and deploy alarm audit Tag objects. The alarm audit Tag objects have attributes that will be bound to the set points, priority and mode of the alarm objects in the network. One alarm audit tag object will be created for every 20 alarms. Below figure 2 shows an overview of the architecture for the system:

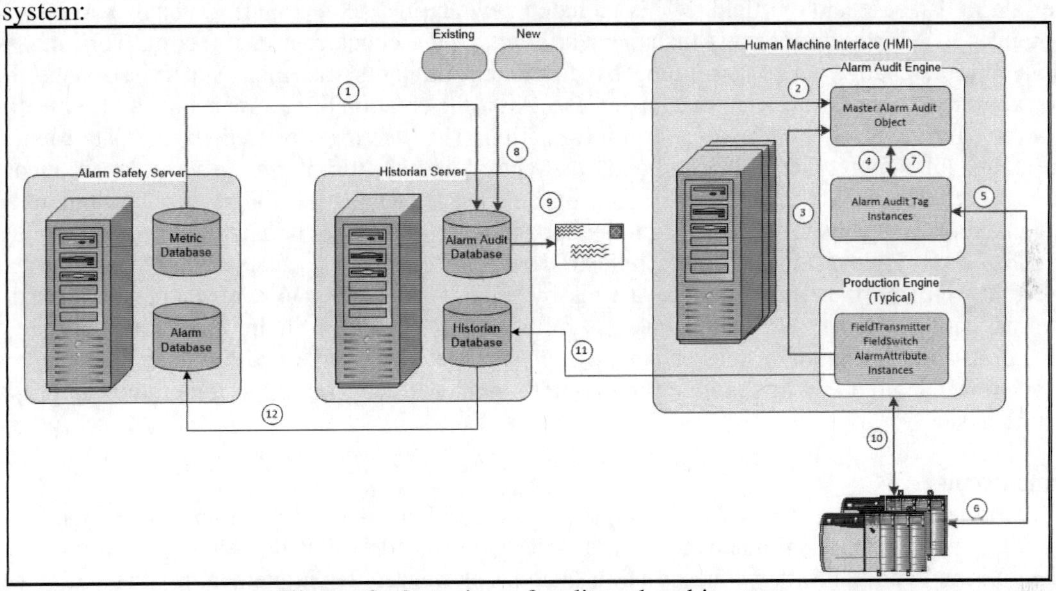

Figure 2: Overview of audit tool architecture

The overall process of the alarm audit and enforcement tool and its interacting with the existing system is listed below that further explains the details and workflow of figure 2:

1. Alarm objective analysis (AOA) info in the Alarm Audit Database is pulled from the metric database by a scheduled job running in Alarm Audit Database
2. AOA info is transferred to Master Alarm Audit Object in HMI
3. Alarm scanner iterates through production field transmitter, field switch, and alarm attributes objects on a regular basis and generates alarm audit tag instances to monitor alarm configuration.
4. Individual alarm configuration data is applied to alarm audit tag instances for comparison and correction.
5. Alarm audit tag instances monitor alarm configuration in HMIs/PLCs on a regular basis.
6. Alarm configuration discrepancies are corrected in PLC(s) when found
7. Alarm audit tag instances report alarm configuration enforcement actions to the Master Alarm Audit Object
8. Alarm configuration audit and enforcement information is transferred back to the Alarm Audit Database
9. A triggered job in the alarm audit database produces email-able reports.
10. Normal PLC-to-HMI SCADA dataflow
11. Normal alarm historization
12. Normal alarm history transfer to alarm database

IOP Conf. Series: Materials Science and Engineering **521** (2019) 012008 doi:10.1088/1757-899X/521/1/012008

The project began with little formal documentation for the alarm master database. By following an agreed upon process, the master alarm database was created. With master alarm database created and containing safety information about each alarm, an automated audit tool was created to cross reference field information with the safety information in the master database. After the system was created and following typical commissioning and start-up checks, the automated audit and enforcement tool went into service.

3. Results

Immediately after formal commissioning and correction of technical issues, the system began generating reports to engineering and management. Reports included any discrepancies between the master alarm database and the field. This included new alarms, deleted alarms, changes to setpoint, and changes to priority. With this management oversite, the concurrent engineering work that was making multiple changes to programming by different contractors was more supervised since there was an automated way to check for discrepancies. After the creation of reports began, alarm meetings also occurred to regularly review the alarm information. The meetings focused on any data found by alarm audit tool, suppression of alarms, and "bad actor" alarms (top alarms in terms of frequency). With this data, operations teams could agree on any needed changes to alarm system and initiate management of change which would lead to an update of the master alarm database. At this time, the system has been deployed to four large industrial sites with five more planned. One of the early adopters has utilized the data to reduce alarms by reclassifying them to a lower priority based on safety analysis. This site has reduced alarms from roughly 55% by utilizing the audit tool with the safety information in the alarm master database. All sites which have had the audit tool deployed have not suffered from any lost production or safety incidents due to multiple programming projects occurring at site.

4. Conclusions

Prior to project, a series of industrial production sites did not have an automated mechanism of reconciling and verifying if implemented alarm settings matched designed settings. Verification and validation is an important safety concept and it includes alarming. By implementing this audit tool, it allowed sites to continue to mature validation and verification of appropriate barriers and safeguards. Alarm configuration changes occur in alarm generating systems sometimes without corresponding changes in the master alarm database. Prior to work, there was no means of validating alarm master database with field alarm configuration systems on a periodic basis. This audit tool bridged this gap. This work led to development of an audit tool that checks for alarm setpoint, alarm priority and alarm HMI tag mismatch but also led to auxiliary improvement in safety. It also checked unauthorized changes to alarm settings. It allowed operations to continually verify if alarms were acting as a protection layer against hazardous events. Giving the operator a reliable alarm configuration information helps the operator maintain plant within safe operating limits and recognize a potentially hazardous condition early and take proper action. Some sites may not have a similar need due to less work activity or less frequent changes to alarms. Need for a similar alarm audit took can vary depending on industry, capital investment, and management controls already in place. There was a large design discussion on the need of continuous vs. batch reports as well. In other words, it was debated if the reports could be generated monthly by a batch scan of data versus instantaneously. This system utilized instantaneous audits, but the decision of how often the reports are generated will also vary based on each application. The result is a system that allows sites to continually verify and validate all facility alarm configurations are safe by use of an automated tool. Future work is looking at ways to further study high frequency alarms against the master alarm database to better rank alarms which need attention. This work is planning to leverage artificial intelligence technology to improve decision making and adding business value for alarm reporting.

5. References

[1] Koene, Johannes, and Hiranmayee Vedam. "Alarm management and rationalization." Third *International conference on loss prevention*. 2000.

[2] Shafer, Don, and Phillip A. Laplante. "The BP Oil Spill: Could software be a culprit?." *IT professional* **12.5** (2010): 6-9.

[3] ANSI-ISA, I. S. A. "18.2 Management of Alarm Systems for the Process Industries." (2009).

[4] Noda, Masaru, et al. "Event correlation analysis for alarm system rationalization." Asia - Pacific *Journal of Chemical Engineering* **6.3** (2011): 497-502.

[5] Hollifield, Bill R., and Eddie Habibi. Alarm Management: *A Comprehensive Guide: Practical and Proven Methods to Optimize the Performance of Alarm Management Systems.* Isa, 2010.

[6] Beebe, Dustin, Steve Ferrer, and Darwin Logerot. "The connection of peak alarm rates to plant incidents and what you can do to minimize." *Process Safety Progress* **32.1** (2013): 72-77.

[7] Vasarhelyi, Miklos A., and Fern B. Halper. "The continuous audit of online systems." *Auditing:* A Journal *of Practice and Theory.* 1991

A Nonlinear Preventive Maintenance Model with an Environmental Factor Based Weibull Distribution

Wenke Gao, Hong Ji, Yu Wang, Tianshu Yang

(School of Energy and Power Engineering, Lanzhou University of Technology, Lanzhou, 730050, China)
E-mail: gaowk_best@163.com

Abstract. Most of the Preventive maintenance (PM) reliability modeling research has focused on the relationship between two adjacent failure intensity functions, while few real case studies have demonstrated the application of these PM models. Thinking from this vision, a nonlinear PM model with scale and shape adjustment parameters is proposed based on a Weibull distribution, and two adjustment parameters can describe each PM effect. Meanwhile, the proposed model can separate the influence of an environmental factor from the failure intensity function of the new system when it operates under a new condition. The environmental parts in the failure intensity function can be improved or removed after several PM actions. Finally, one real case study is exhibited to illustrate the proposed model. The results indicate that the proposed model exhibits good fitting performance in reliability modeling and can describe the PM effect quantitatively as well as reveal the influence of the environmental factor on the system reliability.
Keywords：Nonlinear preventive maintenance model; Weibull distribution; shape adjustment parameter; maintenance effect; environmental factor.

1.Introduction

1.1 Motivation

Preventive maintenance (PM) research, which can increase equipment lifetime and decrease the frequency of in-service breakdowns, has attracted extensive attention since its introduction. According to the maintenance effect, PM can be generally classified into four categories as follows: better-than-perfect PM, perfect PM, imperfect PM and worse PM[1]. A better-than-perfect PM brings a system's operating condition to a state with a smaller failure rate or/and a slower failure process than a brand new identical system. This state can be due to technological advances, more reliable production or performance adjustment in a new environment via maintenance. A perfect PM action restores the system to an "as good as new" condition. Upon perfect maintenance, the failure intensity function of the item is the same as a new one. A worse PM action can increase the failure rate of the item. This type of maintenance might be due to various reasons such as using an inferior lubricant. An imperfect PM action does not lead the system to the "as good as new" or "as bad as old" condition, while it can bring a system to any condition between "as good as new" and "as bad as old" condition[2, 3]. Meanwhile, the environmental influence is a main factor for a new system when it is used in a new condition. This influence may result in some failures when a system operates in its several initial periods, while it can be improved or removed after an operating in the first several PM intervals. Wu and Zuo[2] reviewed the PM policy and pointed out that the existing PM models present weaknesses.

Weibull model has a wide application and some good performances in reliability modeling. As a straightforward analysis approach, Weibull probability plot (WPP) can observe the change in the parameter of Weibull distribution. On the WPP, the monotonicity of the failure intensity function in each PM interval is the same value in the existing PM models, while the real case indicated that it may be different in different PM cycle[4]. Additionally, the environmental factor may play a leading role during the initial operating period when a new system operates in a new condition. These factors can be removed or improved by one or several times the PM after a period of time operating in a new condition, and yet the reliability model cannot be developed using the initial operating data for a new system in a new condition. Otherwise, the system reliability may be misleading. The current PM model was unable to meet these cases, and a few case studies in the literature have demonstrated the use of existing PM models. Based on this consideration, a nonlinear PM model is presented based on Weibull distribution.

1.2 Literature review

As can been seen, the failure intensity function is typically used to describe quantificationally some of these PM effects, as many maintenance actions may not realistically result in a perfect or minimal situation but in an intermediate one[5, 6]. Most research has focused on imperfect PM, whereas there have been few records aimed at better-than-perfect PM. Seo and Bai[7] introduced a periodic PM model. In this paper, the failure intensity function is $h_k(v_{k-1}(t))=h_{k-1}(w_{k-1}(x_{k-2}(t),T))$, where $v_{k-1}(.)$ and $w_{k-1}(.)$ are specified functions, and T is PM interval length. Wu and Derek[8] assumed that the quality of a PM action is a random variable following a probability distribution, and the failure intensity function is $h_k(t)=a^{k-1}h(t)$ after the kth PM, where a is the random value of $a\geq1$. Kijima et al. [9, 10] introduced two types of virtual age PM models, and assumed that PM serves only to remove damage created during the last sojourn, and the virtual age at the start of working after PM is $v_k=t_{k-1}+\xi_k(t_k\ t_{k-1})$ and $v_k=\xi_k(v_{k-1}+\ t_k-t_{k-1})$, where $0<\xi_k<1$. Lam[11], Zhang and Wang[12] et al., have presented the geometric process maintenance. The failure intensity function is $h_k(t)=\alpha h_{k-1}(\alpha t)$ after maintenance. Zequeira and Be´renguer[13] described a system with the following two types of failure mode: maintenance and non-maintenance. A failure intensity function defined as $h_k(t)=\lambda(t)+r(t-(k-1)T)+p_k(t)\lambda(t)$ is introduced, where $\lambda(t)$ is the failure rate of the non-maintenance failure modes, $r(t)$ is the failure rate of the maintenance failure modes and $p_k(t)$ is a function to model the dependence between the maintenance and non-maintenance failure models. Castro[14] also considered a system subject to two modes of failure: maintainable and non-maintainable, while the failure intensity function with definition $h_k(t)=\ \alpha^{N(kT)}h_0(t-kT)$ is related to each failure mode, where $\alpha>1$. Clavareau and Labeau[15] presented a bi-Weibull expression $h(t)=h_0(t)+a(bt-c)^d$ to describe the system failure rate, which allows both the useful period and the ageing zone of a system to be covered with no consideration of infancy problems. Peng et al. [16] proposed a hybrid imperfect maintenance model with random adjustment-reduction parameters and claimed that this model is more realistic in real cases.

The influence of the environment on system reliability is a key factor, while it has been neglected in conventional PM. In practice, the influence of environment may be removed, weakened or improved by several PM actions until it can be ignored in the rest PM intervals. This situation can result in the monotonicity of the system failure intensity function displays different in the remaining PM interval. Wu and Scarf[17] regarded that the failure process of a system may be influenced by operational and environmental stress factors, and described the system failure intensity function with $\lambda_o(t)=\lambda_b(t)\phi(t)$, where is the baseline failure intensity function and is a function of covariates that quantify such extrinsic factors. Xia et al,[18] introduced the effect of an environmental condition by using an environmental factor a_k, where the failure intensity function between the kth and $(k+1)$th PM is $h_{k+1}(t)=a_kb_kh_k(t+c_kT)$. Tao et al,[19] improved Xia's failure intensity model and applied an increase in the hazard rate to describe the environmental factor, and then the model was used to model the reliability of a drilling machine. Xia et al,[20] applied an environment factor to reflect the external environment effect on the machine reliability, and the developed model was used in the predictive maintenance of a manufacturing system. Lhorente et al.[4] provided a case study on wheel motor armatures of a fleet of Komatsu haul trucks in a mining application in Chile. Three successive PM reliability models obtained from real failure data showed that current PM models are unable to represent the real case because the failure intensity function of the 1st PM interval is decreased whereas it increased in the other two PM intervals. Wu and Zuo[2] suggested that PM models maybe consider more complex situations between $h_k(t)$ and $h_{k-1}(t)$.

1.3 Paper outline and definitions

The remaining parts of this paper are arranged as follows: a description of the nonlinear PM model is introduced in section 1, the parameter estimation is presented in section 2, a real case study is examined in section 3, model discussions are offered in section 4, and then a brief summary is given in the last section. Prior to providing a detailed description regarding maintenance policy, some terminologies and definitions used in the forthcoming sections are introduced as follows.
Definitions:
The kth PM restores the system to an "as good as new (AGAN)" state if $h_k(t)=h(t)$.
The kth PM restores the system to an "as bad as old (ABAO)" state if $h_k(t)=h_{k-1}(t_{k-1}+t)$.
The kth PM restores the system to a "better-than-new (BTN)" state if $h_k(t)<h(t)$ almost everywhere.
The kth PM restores the system to a "better-than-old (BTO)" state if $h_k(t)<h_{k-1}(t)$ almost everywhere.

2. Description of the nonlinear PM model

Herein, a general PM policy is considered. Without losing the generality, assume that a PM may be among the BTN, perfect and imperfect states. A minimal repair can be removed, which restores the system to its operating condition if it failed during each PM interval. PM actions are performed at time t_1, t_2,..., t_{N-1}, which can change the healthy condition of the system, and are described by the failure intensity function $h_k(t)$ (k=1,2,...,N). A replacement is employed after (N-1) PMs at t_N, which can restore the system to the state of AGAN. The PM interval between the (k-1)th and the kth PM is x_k that may be a periodic PM, a sequential PM, or a quasi-periodic PM if $x_i=T$, $x_i=T_i$ or x_i is a random variable valued in (T, $T+W$), where T, T_i and the length of maintenance window W are constant[21]. On demonstration of the failure intensity function, the following three points are considered.

a) A non-negligible fact that a failure may be caused in a new system when it operates in a new environment and the environmental factor in the system reliability may be adjusted by some maintenance. This phenomenon, displayed in the failure intensity function, is a competing risk model that includes the influence of both the environmental and the baseline parts. The environmental part can be improved or removed by the early several PM intervals, and the baseline part is related to the PM effect.

b) Another fact is that some systems may experience a periodic PM policy, which may result in a failure intensity function that expresses only a decreasing part in the first several PM intervals. Users may be misled because they only can see the decreasing part of the failure intensity function, and are unable to find the increasing part. Specially, the influence of the environmental part may be weakened by several PM intervals and may be ignored in the remaining PM intervals. Thus, the monotonicity of the system failure intensity function displays differently in the remaining PM interval.

c) The PM effect not only exhibits a change in the scale parameters but also a change in the shape parameters of Weibull distribution in different PM intervals.

Based on the above consideration, the following assumptions are used to develop the proposed model.

1) The lifetime distribution of the system follows Weibull distribution, where the baseline failure intensity function $h_B(t)$ is a two-parameter Weibull distribution. The environmental influence on the system reliability decreases with the system operating time, and can be described by a decreasing function $h_Q(t)$ that is a two-parameter Weibull model in the failure intensity function. Thus, the failure intensity function for the new system is $h(t)$.

$$\begin{aligned} h(t) &= h_B(t) + h_Q(t) \\ &= \frac{b}{h}\left(\frac{t}{h}\right)^{b-1} + \frac{a}{q}\left(\frac{t}{q}\right)^{a-1} \end{aligned} \tag{1}$$

where η and θ denote the scale parameters and $\beta(\beta>1)$ and $\alpha(\alpha<1)$ denote the shape parameter in a new system.

2) The failure intensity function of the system after the kth PM is $h_k(t)$, which meets the following relationship,

$$h_k(t) = A_k h_B(t^{B_k}) + C_k h_Q(t^{D_k}) \tag{2}$$

where $A_k=a_1a_2\ldots a_k$, $B_k=b_1b_2\ldots b_k$, and $t\geq 0$ is working time after the kth PM. The parameters a_k, c_k are the scale adjustment parameters after the kth PM, and a_k, c_k >0. The parameters b_k, d_k are the shape adjustment parameters after the kth PM, and b_k, d_k >0. The environmental part can be improved, removed or even ignored when the system experiences k PMs, and the failure intensity function can be stated as follows,

$$h_{k+m}(t) = A_{k+m} h_B(t^{B_{k+m}}) \tag{3}$$

3) All failures can be instantly detected and repaired. Each PM can bring the system to any state among the better-than-perfect, perfect or imperfect states. The system is restored to AGAN at replacement. The times for minimal repairs, PMs and replacements are negligible.

4) The mean cost of minor repair C_f is unrelated to the occurred failures and the total severed times of the system. The cost of PM C_p is not relevant to the total operational time of the system. The cost of a replacement is a constant C_r (C_r>C_m).

According to the above description of the maintenance process and model assumptions, an optimization model of the proposed imperfect sequential PM policy is presented. The system long-running cost rate is stated as below,

$$g(\{T_N\}, N) = \frac{C_p(N-1) + C_f \sum_{k=1}^{N} H_k(T_k) + C_r}{\sum_{k=1}^{N} T_k} \tag{4}$$

The optimal value ($\{T_N^*\}$, N^*) that can minimize the function $g(\{T_N\};N)$ is given in Eq. (4). A unique optimal sequential PM policy with $\{T_N^*\}$ and N^* may be found under a consideration of A_k and B_k [21].

3. Parameter estimation

For a lifetime cycle, the models include a new system and preventively maintained system. Similarly, parameter estimations also involve α, θ, η, β, a_k, b_k, c_k and d_k, and then parameters A_k, B_k, C_k and D_k can be obtained according to definition of a_k, b_k, c_k and d_k. For failure data, complete data and censor data may exist in a PM interval, and thus, these two types of failure data are considered in the following reliability modeling.

Here, Let t_i and τ_i denote the failure time and the censored time for the observed individual i, i = 1, …, n+m. For complete data, m=0. If t_i and τ_i are independent random variables, then the log-likelihood function

$$L(t\,|\,X) = \prod_{j=1}^{n} f_k(t_j\,|\,X) \prod_{j=1}^{m} R_k(\tau_j\,|\,X) \tag{5}$$

$$L(t\,|\,X) = \prod_{j=1}^{n-m} (A_k h_Q(t_j^{B_k}\,|\,X) + C_k h_B(t_j^{D_k}\,|\,X)) e^{(-A_k H_Q(t_j^{B_k}|X) - C_k H_B(t_j^{D_k}|X))} \prod_{j=1}^{m} e^{(-A_k H_Q(\tau_j^{B_k}|X) - C_k H_B(\tau_j^{D_k}|X))} \tag{6}$$

If only the failed items are included, then the log-likelihood function ln$L(t|X)$ can be described as,

$$\ln L(t\,|\,X) = \sum_{j=1}^{n-m} \left(\ln(A_k h_Q(t_j^{B_k}\,|\,X) + C_k h_B(t_j^{D_k}\,|\,X)) \right) - \sum_{j=1}^{n} \left(A_k H_Q(t_j^{B_k}\,|\,X) + C_k H_B(t_j^{D_k}\,|\,X) \right) \tag{7}$$

where, $0<\alpha<1$ *and* $\beta>1$

Setting the first partial derivatives of $\ln L(t|X)$ with respect to each parameter to zero. By solving the systems of the nonlinear likelihood equation, and then we can obtain the maximum likelihood estimates for the complete and censored data.

4. Real Case Study

A study in reference[4] explored wheel motor armatures of a fleet of Komatsu haul trucks in a mining application in Chile. In this real case, four years of maintenance data of these components were analysed. The failure data was separated into three groups according to PM frequencies: new armatures (before the first PM), after the first but before the second PM, and after the second PM. In the paper, two-parameter Weibull models are obtained for each PM cycle. The shape and scale parameters are: $\beta_{1\text{-}3}$ (0.809, 1.247 and 1.235), $\eta_{1\text{-}3}$ (36286, 6382 and 5264).

It can be found that the authors were unable to study the relationship among the three models, and the shape parameters from β_1 to β_3 increased with the PM frequencies. During the first operating interval, the system is new with $\beta_1<1$, which means that the system exhibited a decreasing failure intensity with operating time before the first PM is performed. This was not always the case and maybe caused by environmental or production quality, and the first PM was performed so early that the whole trend of the failure intensive function was unable to show completely. Obviously, the whole failure intensity function should be a "bathtub" type. This case is suitable to illustrate the proposed nonlinear PM model.

To develop the quantitative relation between adjacent PM cycles, two data sets are created from the presented reliability model of the investigation[4] because the failure data was not exhibited in the paper. One data set is a complete data with the 200 group data, and the other is a censor data also with the 200 group data. Consequently, the modeling process and maintenance optimization are given as follows.

4.1Reliability Modeling

Step 1. Create $\{R_i(t_j)\}$

Twenty groups of random data are created within [0.8,1] and are viewed as reliability data sets for the new system using the new system parameters. For the complete data, each data set has 50 failure data, and it has failure data and non-failure data with the total number of 50 for the censor data. Similarly, 200 groups of random data are created within [0,1] and are viewed as reliability data sets for the second and third PM interval. Each group has 100 failure data, which are then marked as $\{R_i(t_j)\}$, where i=1,2,3 denotes the PM interval and j=1,2,...50 or 100.

Step 2. Obtain the failure data $\{t_i\}$

Failure data sets $\{t_i\}$ are attained according to the created $\{R_i(t_j)\}$ and the two-parameter Weibull distribution function of each PM cycle.

Step 3. Estimate the parameters

Using the parameters estimation methods to obtain each parameter, and then compute their mean values as the system parameters. The parameters for complete data and censored data are obtained as bellows.

For the new system, α=0.7837, θ=40629, β=1.9742, η=59695. After the 1st PM, a_1=30497 and b_1=0.2535. After the second PM a_2=1.3956 and b_2=0.9514.The new system is a four-parameter Weibull model. After the two PM actions, the failure intensity is changed. It can be found that the environmental factors are removed after operating in the first PM interval. The parameters a_2 and b_2 can exhibit the second PM effect. β and η can reveal the system reliability in the new system, and θ and α display the environmental factors.

4.2 Maintenance optimization

In this case, a sequential PM policy is considered and takes C_p=20, C_f=40 and C_r=60, where $\alpha<1$ and $\beta>1$, that is $h(t)$ is a "bathtub" type function. According to Theorem 4, there are two solutions. The first solution is near zero and is ignored, and the optimal $\{T_i^*\}$ exists. Then, it can be found that the optimal $N^*=1$ and $T_1=27414$. The results show that the sequential PM or PM policy is unnecessary for the system, while the periodic replacement policy is optimal.

4.3 Comparison with the linear PM model

Based on the obtained failure data, the differences between the proposed PM model and traditional linear PM model are analysed here, which only consider one linear PM model with failure intensity function $h_k(t)=a_k h(t)$. For the new system, the reliability function is the same as the above. After the 1st PM but before the second PM, the function has one unknown parameter a_1 that can be attained through the MLE method. Similarly, a_2 can also be obtained. The failure intensity function curves of the proposed nonlinear and linear PM model are shown in Fig.1.

From the Fig. 1, the proposed PM model has a better fit than the linear PM model. To evaluate the goodness-of-fit, the BIC value is usually utilized. Based on the estimated linear PM model and the proposed model, the BIC values are obtained. After the 1st PM, the BIC values of the proposed model and the linear PM model with the complete data are 1946 and 2290. After the 2nd PM, the BIC values are 1908 and 2333. For the censored data, the BIC values are 1946, 1970, and 1908, 1955 respectively. Obviously, the BIC value of the proposed model is smaller, and thus the proposed PM model has a greater goodness-of-fit than the linear PM model.

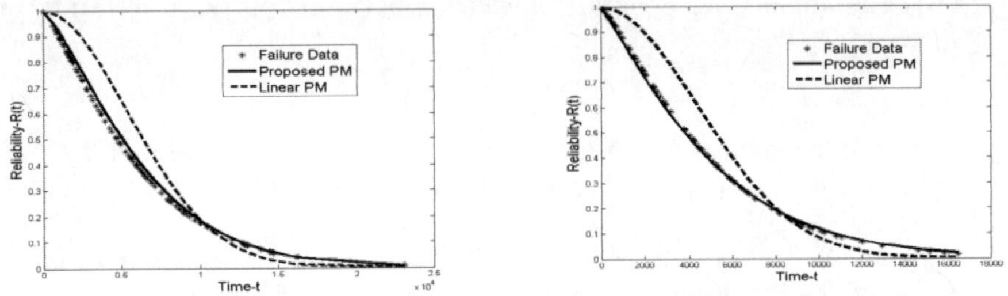

a) After the 1st PM b)After the 2nd PM

Fig. 1 $R(t)$ with different models

4.4 Environmental factors analysis

The influence of environmental factor on the system failure intensity requires more attention during the initial several PM intervals for a new system and can be removed by an operating adjustment or maintenance. Together with the system baseline failure intensive factor, the system reliability can be seen as a competing risk model. With the growth of the system operating time, the environmental factor can be weakened until it is ignored, while the system baseline failure intensive factor may be heightened. Herein, we illustrate this point using the proposed model.

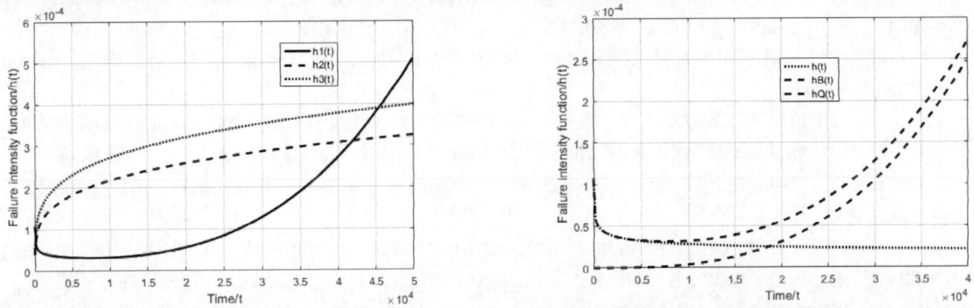

a)Failure intensive function for 3 PM intervals b)Failure intensive function for the new system

Fig. 2 Failure intensive function

In the subplot of Fig. 2, $a_k>1$ and $b_k<1$. It can be found that the initial value of the environmental factor in the failure intensity function in the 2nd PM interval can be ignored because the time for a large value is transient and then approach zero.

5. Results

In this paper, a nonlinear PM model $A_k h_B(t^{Bk})+C_k h_Q(t^{Dk})$ is proposed based on the Weibull distribution, and parameter estimation methods are given. The proposed model not only depicts the effect of some PM actions that may restore the system to any condition between AGAN and BTO, but also may bring the system to BTN or even to ABAO. Specially, the model can provide a goodness-of-fit for the failure data in reality. These properties exist in practice and are mainly described by parameter A_k and B_k. Additionally, the proposed model can describe the influence of the environmental factors on the system reliability in different PM interval via parameters C_k and D_k. The influence can be improved or removed by maintenance actions in the first several PM intervals. The property is proved by the real case study. The existence condition of the optimal periodic PM policy exhibited that the optimal PM policy does not always exist, which is decided by A_k, B_k, $h(t)$ and the maintenance cost of each item. A real case study is given to support the model's application and properties.

Acknowledgments

The research work is also supported jointly by the National Natural Science Foundation of China (grant number 71561016) and China Postdoctoral Science Foundation (grant number 2017M613297XB). The authors are grateful for valuable supports of Wu Centre of University of Michigan.

Reference

[1] H P, H W. Imperfect maintenance. European Journal of Operational Research. 1996;94:438-52.
[2] Wu S, Zuo MJ. Linear and Nonlinear Preventive Maintenance Models. IEEE TRANSACTIONS ON RELIABILITY. 2010;59:242-9.
[3] Doyen L, Gaudoin O. Classes of imperfect repair models based on reduction of failure intensity or virtual age. Reliability Engineering and System Safety. 2004;84:45-56.
[4] Lhorente B, Lugtigheid D, Knights PF, Santana A. A model for optimal armature maintenance in electric haul truck wheel motors: a case study. Reliability Engineering and System Safety. 2004;84:209-18.
[5] Basri EI, Razak IHA, Ab-Samat H, Kamaruddin S. Preventive maintenance (PM) planning: a review. Journal of Quality in Maintenance Engineering. 2017;23:114-43.
[6] Nakagawa T, Mizutani S. A summary of maintenance policies for a finite interval. Reliability Engineering and System Safety. 2009;94:89-96.
[7] Seo J, Bai D. An optimal maintenance policy for a system under periodic overhaul. Mathematical and Computer Modelling. 2004;39:373-80.
[8] Wu S, Clements-Croome D. Preventive maintenance models with random maintenance quality. Reliability Engineering and System Safety. 2005;90:99-105.
[9] M K. Some results for repairable systems with general repair. Journal of Applied Probability. 1989;26:89-102.
[10] Kijima M, Morimura H, Suzuki Y. Periodical replacement problem without assuming minimal repair. European Journal of Operational Research. 1988;37:194-203.
[11] Y. L. Geometric processes and replacement problem. Acta Mathematicae Applicate Sinica. 1988;4:366- 77.
[12] Zhang Y, Wang G. A geometric process repair model for a repairable cold standby system with priority in use and repair. Reliability Engineering and System Safety. 2009;94:1782-7.
[13] Zequeira RI, Be´renguer C. Periodic imperfect preventive maintenance with two categories of competing failure modes. Reliability Engineering and System Safety. 2006;91:460-8.
[14] Castro IT. A model of imperfect preventive maintenance with dependent failure modes. European Journal of Operational Research. 2009;196:217-24.
[15] Clavareau J, Labeau PE. An alternative imperfect preventive maintenance model2009.

[16] Peng W, Liu Y, Zhang XL, Huang HZ. Sequential preventive maintenance policies with consideration of random adjustment-reduction features. Eksploat Niezawodn. 2015;17:306-13.
[17] Wu S, Scarf P. Decline and repair, and covariate effects. European Journal of Operational Research. 2015;244:219-26.
[18] Xia T, Xi L, Lee J, Zhou X. Optimal CBPM policy considering maintenance effects and environmental condition. Int J Adv Manuf Technol. 2011;56(9):1181–93.
[19] Tao X, Xia T, Xi L. Environment factor–based equipment hazard rate prognosis for maintenance scheduling. Proceedings of the Institution of Mechanical Engineers, Part B: Journal of Engineering Manufacture. 2017;231:1488-97.
[20] Xia T-B, Tao X-Y, Xi L-F. Operation process rebuilding (OPR)-oriented maintenance policy for changeable system structures. IEEE Transactions on Automation Science and Engineering. 2017;14:139-48.
[21] Gao W, Zhang Z, Ji H, Zhou Y, Liu Q. Optimal Quasi-Periodic Preventive Maintenance Policies for Repairable System with Dynamic Maintenance Plan. Eksploat Niezawodn. 2015;17:389-97.

IOP Conf. Series: Materials Science and Engineering **521** (2019) 012010 doi:10.1088/1757-899X/521/1/012010

Application of Technological System Based on Processes Modelling and Analysis: A Case Study in Italian Aquaculture Company

Angelo Corallo[1], Maria Elena Latino[1], Marta Menegoli[2], Marco Cataldo[3], Luisa Mancarella[3]

[1] University of Salento, Department of Innovation Engineering, Edificio Aldo Romano, Campus Ecotekne, Via per Monteroni s.n., 73100 Lecce, Italy. angelo.corallo@unisalento.it, mariaelena.latino@unisalento.it

[2] Naica sc, Edificio Dhitech, Campus Ecotekne, Via per Monteroni 165, 73100 Lecce, Italy. martamenegoli@naicasc.com

[3] Apphia s.r.l., Edificio Dhitech, Campus Ecotekne, Via per Monteroni 165, 73100 Lecce, Italy. marco.cataldo@apphia.it, luisa.mancarella@apphia.it,

Corresponding author's e-mail: mariaelena.latino@unisalento.it

Abstract. Aquaculture is a growing sector and could satisfy the future worldwide fish demand. New sustainable consumption patterns arise from food market, based on the consumer' need to know more information about food product. Technological and methodological innovations are needed to create more efficiency in aquaculture industry, but also to create product able to respond to the emerging market needs. The paper shows a case study in Aquaculture Company useful to demonstrate how collect product and production processes information and use it for traceability purpose. Starting from the as-is production process analysis and modelling, several strategically information for consumer are identified. The possibility of using the devices developed in S.E.A. research project has been exploited to collect some of this information and a new traceability layer was create in order to collect manually the missing data. At last the new production process was modelled. The benefits coming from the system adoption are multiple, for the final consumer that could see demonstrated the food quality and safety, for the company that could increase own environmental, economical and ethical sustainability.

1. Introduction

According to Food and Agricultural Organization (FAO) forecasts, in 2030 the demand for fish will be about 261,2 million tons. In order to meet this worldwide demand, therefore, the 62% of fish products will be produced using aquaculture techniques, as an alternative to wild fisheries captured [1]. So, it is necessary take into account several issues of aquaculture industry improving: quality and safety of fish, health and well-belling of animals during their life cycle, operators' safety during production activities and aquaculture sustainability from environmental, economic and social [2] points of view. From literature background arose as sustainable production methods are strategic issue also for consumers [3,4]. Implementing sustainable practices could provide an opportunity to differentiate the companies' value proposition [5]. Sustainable aquaculture is assumed to match the demand of a particular consumer segment that appreciates additional environmental and ethical values of products [6–8]. More recently, in addition to sustainability issues, concerns about animal welfare are becoming more relevant in public opinion and also in consumers' food demand [9,10].

Taking account this scenario, the purpose of the study was to propose a new concept of integrated smart system able to supply to the final consumer detailed information about fish produced in aquaculture plant. So, the following central Research Question (RQ) guides the study:
RQ – How to collect the data coming from the fish production process in order to satisfy new emerging food traceability demand of the consumer?
The present work represents a follow-up of a Research & Development project accomplished in Italian country (Security for marine Environment and Aquaculture, S.E.A. - www.clustersea.it) that aimed to foster the environmental and economic sustainability of Aquaculture Industry through the development of an innovative solution for aquaculture plant. The S.E.A. devices (described in [11]) were perfectly integrate with existing structures and allow the company to obtain data about production. The elaboration of this data could provide different type of information. The challenge consists into recovery the information from different activities, actors and device and makes it shareable with final consumer, following the food traceability point of view.

2. Materials and methods

In order to answer to the new emergent consumer needs about sustainability and ethical values in food production and leveraging on S.E.A. system, the case study method [12,13] was chosen analysing the operative scenario of the aquaculture company involved in S.E.A. as partner. This qualitative method is able to examine contemporary real-life situations and provide the basis for the application of ideas and extension of method [14]. The purpose of the case study is to analyse, from one side, the company' production process in order to understand in which activity/actor/device resides the information needed by the market and, on the other side, evaluate how the technology already developed in S.E.A. can provide and make shareable this information. The tool used to realize the case study was the qualitative face-to-face interview [13]. The interview was based in a set of questions useful to investigate how the employees operate during daily working activities and was completed by the following information: time, data and place of interview, company role of the interviewee. The interview protocol foresaw that the researchers record information by audiotaping, so the answers was transcript and used to model business process. The interviewees hold four different roles (Plant Manager, Production Operator, Maintenance Operator and Diver) and provide the information useful to model the fish production process using Business Process Model Notation (BPMN) methodology and Signavio tool (www.signavio.com).

3. Collect fish traceability information during aquaculture production, case study

From literature analysis arise as consumer wants to know more information about food product (see [15]). Three main categories of information were taken into account for Aquaculture Industry: information about the product (P), information about the production environment (PE) and information about the company (C). **Table 1** summarizes a short description of the S.E.A. devices, the data that each S.E.A. device is able to collect and the related category of information. For more details about S.E.A. devices see [11].

Table 1. Traceability data collected automatically by S.E.A. devices.

S.E.A. Device	*Data*	*Categories of information*
Device for biomass forecast - based on stereoscopic vision, the device realizes a non-invasive statistical measure. It uses a set of algorithms of Computer Vision science and the length-weight parameters for the fish, in order to monitor fish growth curve. The device aims to define a right food quantity to give, related to the fish growth level, reducing the wastefulness and increasing the production quality.	Fish average weight	P
	Fish growth rate	P

Device for marine data collection and forecast - it is composed by several tools such as buoys and sensors, that capture the information, and by a software module able to analyze the data and apply forecast models. It measures waves and currents, and other relevant variables (e.g. chlorophyll, phytoplankton, dissolved molecular oxygen and nitrate/phosphate concentration) are modelled at different depths in order to identify potential biogeochemical alteration around the cages. The models are validating through the use of satellite and in-situ observations.	Water temperature	PE
	Cage Position	PE
	PH	PE
	Water turbidity	PE
	Salinity water	PE
	Oxygen dissolved in H_20	PE
Device for fish farm cages monitoring - it is composed by an underwater video camera and a web application for video stream management, in order to identify and select in real-time interesting frame about an anomaly of the net and its GPS position.	N ° net damages	PE
Device for predators' detection - equipped with a hydrophone, like an electronic ear, it is able to identify the presence of a predator around the plant and to give an alert to the Plant Manager in order to perform the removal activities.	N ° predators alarms	P
Device for solid waste recovery - it is a collector positioned at the bottom of the cage net. The waste particles accumulated by gravity, are periodically removed through a flexible tube connected with a dewatering system. Here, the solid wastewater will pass in a mechanical filter. The filtered water will be pumped out to the sea while the concentrated sludge will be re-use and/or disposal. Through this device the pollution in the plant area will be reduced.	Kg of recovered solid waste	PE

However, in order to better fulfil the central research question, other information about the product was considered for each category: i) Product: commercial denomination, scientific name, production method, FAO code, nutritional values, presentation, state of transformation, freshness, destination and the codes required by European standards, Batch number, operator names, dates of sowing and fishing, Type and amount of feed used, Curiosity about the product; ii) Company: country of production, company demographic information; iii) Production environment: breeding habitat description.
The production process, using S.E.A. devices, was modelled in order to demonstrate how collect traceability information about food as shown in **Figure 1**, where the dark grey activities are those supported by the proposed system. Downstream the process modelling, it made possible to identify for each activity the data relevant for traceability of farmed product. **Table *2*** summarizes for each activity, the mapped information, the source of the data and the collection method used (manual data entry or automatically data entry). A module of Information Technology (IT) platform, dedicated to traceability, was developed in order to allow manual data entry of the missing information.

4. Conclusion and follow-up

From European and worldwide scenario arose as Aquaculture Industry has the potential to grow in the next years, in order to satisfy the global fish demand. At the same time, plant innovation could take into account environmental and ethical impacts of production and the increasing care of the consumer in health and wellness, that ask more information about food product. The S.E.A. project proposed a system composed by several Internet of Things (IoT) devices and an IT platform able to manage plant information, in order to assure an innovative and sustainable approach in aquaculture plant management. This information can be used for traceability purpose in order to demonstrate to final consumer the food quality and safety. Using the S.E.A. system and the traceability approach the company could increase:

IOP Conf. Series: Materials Science and Engineering **521** (2019) 012010 doi:10.1088/1757-899X/521/1/012010

i) environmental sustainability, allowing to monitor environmental impact coming from the production; ii) economical sustainability, allowing to produce a quality product for higher market segment; iii) ethical sustainability, allowing to be transparent for the consumer. To date, the system satisfies the food traceability demand of the consumers through data collected automatically by S.E.A. devices and some data manually inserted by the operator in a dedicated module of the IT Platform. In order to improve it and take advantage of the related benefits, next step is the finalization of the traceability information by an innovative and automatically way to collect information actually inserted by manual data entry and to provide to the customer the visualization of the tracking data via the Quick Response (QR) code, using the smartphone.

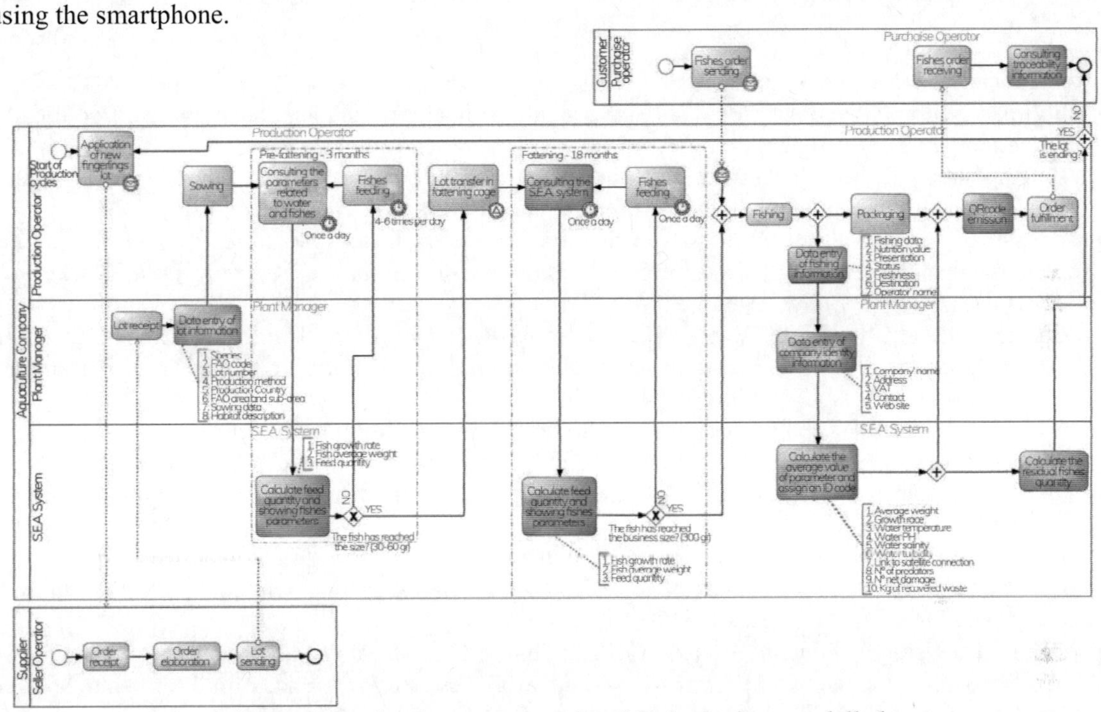

Figure 1. Production process using S.E.A. devices modelled.

Table 2. All traceability information mapped in the activities within the new Production Process.

Activities	*Data*	*Data source*	*Collection methods*
Data entry of lot information	Species, FAO code, lot number, production method, production country, FAO area and sub-area, sowing date, habitat description.	Plant Manager	Manual data entry
Calculate feed quantity and showing fishes parameters	Fish growth rate, fish average weight, feed quantity.	S.E.A. System	Automatically data entry
Calculate feed quantity and showing fishes parameters	Fish growth rate, fish average weight, feed quantity.	Plant Manager	Manual data entry
Data entry of fishing information	Fishing date, nutritional values, presentation, status, freshness, destination, operator' name	Production Operator	Manual data entry

Data entry of company identity information	Company' name, address, VAT, contact, website.	Plant Manager	Manual data entry
Calculate the average value of parameter and assign an ID code	Average weight, growth rate, water temperature, water PH, water salinity, water turbidity, oxygen in water, position, link to satellite connection, n° of predators, n° net damage, kg of recovered waste.	S.E.A. System	Automatically data entry

5. References

[1] Soliman N F and Yacout D M M 2016 Aquaculture in Egypt: status, constraints and potentials *Aquac. Int.* **24** 1201–27

[2] Barrington K, Ridler N, Chopin T, Robinson S and Robinson B 2010 Social aspects of the sustainability of integrated multi-trophic aquaculture *Aquac. Int.* **18** 201–11

[3] Bergleiter S and Meisch S 2015 Certification Standards for Aquaculture Products: Bringing Together the Values of Producers and Consumers in Globalised Organic Food Markets *J. Agric. Environ. Ethics* **28** 553–69

[4] Carlucci D, Nocella G, De Devitiis B, Viscecchia R, Bimbo F and Nardone G 2015 Consumer purchasing behaviour towards fish and seafood products. Patterns and insights from a sample of international studies *Appetite* **84** 212–27

[5] Osterwalder A 2004 The business model ontology: A proposition in a design science approach

[6] Altintzoglou T, Verbeke W, Vanhonacker F and Luten J 2010 The Image of Fish from Aquaculture Among Europeans: Impact of Exposure to Balanced Information *J. Aquat. Food Prod. Technol.* **19** 103–19

[7] Kalshoven K and Meijboom F L B 2013 Sustainability at the Crossroads of Fish Consumption and Production Ethical Dilemmas of Fish Buyers at Retail Organizations in The Netherlands *J. Agric. Environ. Ethics* **26** 101–17

[8] Zander K, Risius A, Feucht Y, Janssen M and Hamm U 2018 Sustainable Aquaculture Products: Implications of Consumer Awareness and of Consumer Preferences for Promising Market Communication in Germany *J. Aquat. Food Prod. Technol.* **27** 5–20

[9] Broom D M 2010 Animal Welfare: An Aspect of Care, Sustainability, and Food Quality Required by the Public *J. Vet. Med. Educ.* **37** 83–8

[10] Weible D, Christoph-Schulz I, Salamon P and Zander K 2016 Citizens' perception of modern pig production in Germany: a mixed-method research approach *Br. Food J.* **118** 2014–32

[11] Corallo A, Errico F, Latino M E, Menegoli M, Calabrese F, Cataldo M and Zizzari A A 2018 Advanced system for sustainable aquaculture plant management *2018 7th International Conference on Industrial Technology and Management (ICITM)* 2018 7th International Conference on Industrial Technology and Management (ICITM) (Oxford, United Kingdom: IEEE) pp 162–6

[12] Yin R K 1994 Case study: design and methods *Newbury Park Sage Publ.*

[13] Creswell J W and Creswell J D 2017 *Research design: Qualitative, quantitative, and mixed methods approaches* (Sage publications)

[14] Soy S 2015 The case study as a research method

[15] Latino M E, Fortunato L, Menegoli M, Scarafile G, Errico F and Corallo A 2018 Ethical design in ICT application: how satisfy food citizenship needs *Proceedings of the 2nd International Conference on High Performance Compilation, Computing and Communications - HP3C* the 2nd International Conference (Hong Kong, Hong Kong: ACM Press) pp 73–7

IOP Conf. Series: Materials Science and Engineering **521** (2019) 012011 doi:10.1088/1757-899X/521/1/012011

Experiment and simulation on heavy oil production with steam flooding in heterogeneous reservoir

K Maneeintr[1], K Sasaki[2] and T Boonpramote[1]

1 Carbon Capture, Storage and Utilization Research Group, Department of Mining and Petroleum Engineering, Faculty of Engineering, Chulalongkorn University, Bangkok 10330, Thailand
2 Department of Earth Resources Engineering, Faculty of Engineering, Kyushu University, Fukuoka, Japan

Corresponding author: Krengkrai.M@chula.ac.th

Abstract. Heavy oil has long been known as one of the main energy sources due to the huge amount of reserves. Heavy oil production depends on reservoir characteristics like permeability, reservoir pressure as well as oil viscosity. The difficulty to produce this oil is its high viscosity. The practical technology to produce this heavy oil is to use steam as steam-flooding method to reduce oil viscosity. However, the huge energy consumption in steam-flooding operation can impact the operating cost as well as a project life. Therefore, the objectives of this study are to measure the viscosity and correlate the results with temperature and to simulate the oil production by using STARS, a CMG program. Moreover, the effects of parameters like well distance, injection rate and permeability on the oil production are also investigated. The simulation results show that oil viscosity and permeability play significant roles in heavy oil production. Also, the higher oil recovery can be obtained by increasing the steam injection rate and shorten the well spacing. These conditions can be applied to use in the real field for heavy oil production. The major benefit of this practice is to reduce the steam consumption and fuel costs and increase more oil production thus extending the economic project life.

1. Introduction

Heavy oil has long been known as one of the main energy sources because of the huge amount of reserves and higher demand of oil consumption [1]. Heavy oil production depends on reservoir characteristics like permeability, reservoir pressure as well as oil viscosity. However, with the high viscosity of heavy oil, the practical technology to produce this heavy oil is steam-flooding method, one of the main techniques for enhanced oil recovery [2-3]. In fact, the steam-flooding performance can be linked to several design parameters such as oil recovery, well spacing, steam injection rate and amount of steam used. Steam consumption per oil production or cumulative steam-oil ratio (Cum SOR) can be used as a criteria to consider the steam-flooding method. Also, to determine the favorable operating conditions, many recent researches have been studied [4-5]. Therefore, the objectives of this study are to measure the viscosity and correlate the results with temperature and to simulate the oil production by using STARS, a CMG program with the five-spot pattern. Furthermore, the effects of parameters such as well distance, injection rate and permeability on the oil production are also investigated. These operating conditions can be applied to use in the real field for heavy oil production. The major benefit of this practice is to reduce the amount of steam consumption and fuel costs and to increase more oil production. Therefore, the project life of oil production can be extended and economically viable.

2. Viscosity measurement

2.1 Equipment and procedure

Oil viscosity and its emulsions are measured with the effect of temperature at 6.58 and 3.20 oil/water ratio. The measurement temperature ranges from 15 to 180°C. The viscosity measurement is carried out by Brookfield cone/plate programmable viscometer Model DV-ПI, using a cone spindle with various numbers. The viscosity measurement is performed with the accuracy of ± 1.0ºC. The accuracy of viscosity measurement is ± 1.0% accuracy of full scale range for a specific spindle running at a specific speed. Allow enough time for the sample to reach equilibrium. Each sample can be measured 3 times to get the average result. Calibration can be done with standard solution supplied by Brookfield.

2.2 Experimental results

Heavy oil and its emulsions are measured at temperatures from 15-180°C as presented in Figure 1. From the results, it is obvious that the viscosities decrease as the temperature increases because of thermal energy. The viscosities of the mixture of oil and steam or emulsions with 6.58 and 3.20 oil/water ratio (O/W) are higher than that of original oil for the whole concentration range.

These phenomena can be explained by the fact that crude oil, especially the heavy oil, contains the large quantities of asphaltenes (high molecular weight polar components) that act as natural emulsifiers. Other crude oil components are also surface active: resins, fatty acids such as naphthenic acids, wax crystals, etc. These substances contain hydrogen bonds in the heavy oil molecule. When steam is present in heavy oil, hydrogen bonds are formed following interactions between components and water. Hydrogen bonds can create irregular properties deviated from mixing rules such as maximum or minimum peak for viscosity of mixture [6-7].

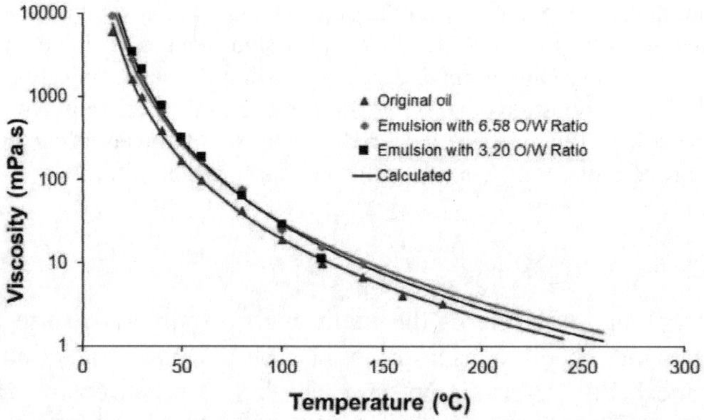

Figure 1. Effect of temperature on viscosities of oil and its emulsions

3. Correlation for viscosity measurement

The correlation of viscosity as a function of temperature to relate the experimental data is developed based on the ones from the literature [7-9] as presented in equation 1. Therefore, as shown in Figure 1, the calculated results from correlation are calculated and compared with the experimental results. This correlation is in the form as shown below:

$$\mu = A\ T^{-B} \tag{1}$$

Where μ = viscosity (cP),

IOP Conf. Series: Materials Science and Engineering **521** (2019) 012011 doi:10.1088/1757-899X/521/1/012011

T = temperature (°C)
A and B = coefficients

In addition, the average absolute deviations (AADs) between experimental and calculated values is 12.92 %. Therefore, this correlation can be acceptable to predict the viscosity of heavy oil.

4. Simulation

4.1 Reservoir model

The simulation study is performed by using STARS program from Computer Modeling Group (CMG). The reservoir model is constructed from the actual field data such as permeability ranging from 1,000 to 25,000 mD. The inverted 5-spot pattern [10] which has an injection well at the center and 4 production wells at the corners with 80 % steam quality is used in this study as illustrated in Figure 2. The three different well distance from 150 to 250 m are evaluated by considering the production efficiency and recovery factor (RF) with various injection rates ranging from 20-100 m^3/d.

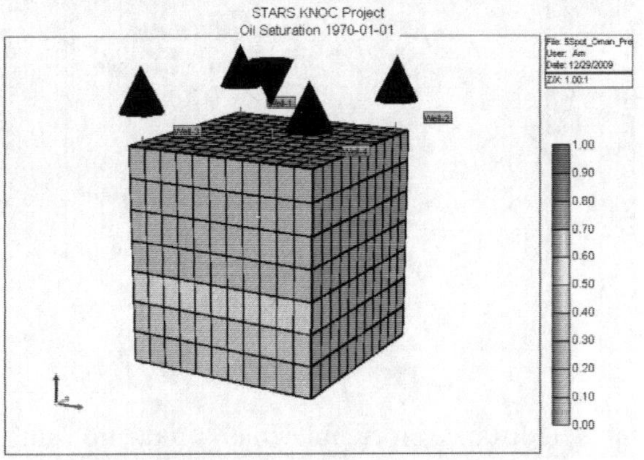

Figure 2. Reservoir model simulation

4.2 Simulation results

4.2.1 Effect of permeability on oil production. Practically, the permeability of the well can be varied randomly from low to high. However, to investigate the effect of permeability, the average permeabilities are estimated at 1000, 5000 and 25000 mD which are the average of the field data for each area. The effect of permeability on the percent recovery factor (%RF) and cumulative steam oil ratio (SOR) is shown in Figure 3.

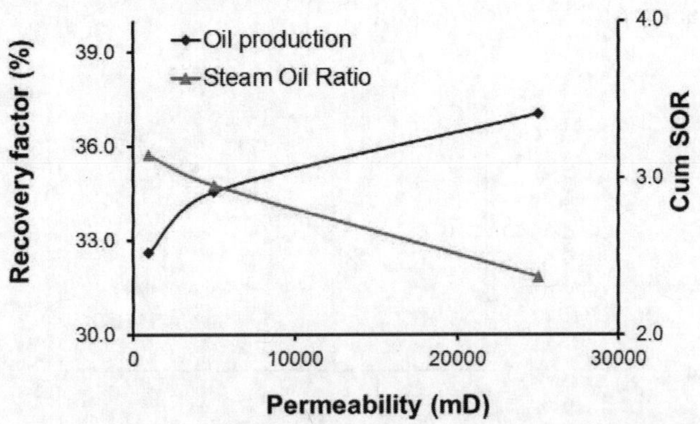

Figure 3. Effect of permeability on cumulated oil production and steam oil ratio

It is relatively clear that at high permeability, the oil production is higher compared to that at low permeability because the fluids can flow easily and steam injection rate can be reduced. This condition is conducive to heavy oil production. From SOR point of view, it can be reduced significantly as permeability increased especially from 5000 mD to 25000 mD. This means less amount of steam is needed to produce heavy oil. Compared to 5000 mD, the cumulative oil productions at 1000 and 25000 mD are 8.19 % lower and 26.59 % higher, respectively and consumed SOR are 6.47 % higher at low permeability and 19.64 % lower at high permeability.

4.2.2 Effect of well distance on oil production. The results of the effect of well distance on oil production is presented in Figure 4. From the results of three different injector-producer distances, 150, 200 and 250 m, it presents that the shorter distance provides higher oil recovery than of the longer one. However, the cumulative SOR is decreasing with an increasing distance.

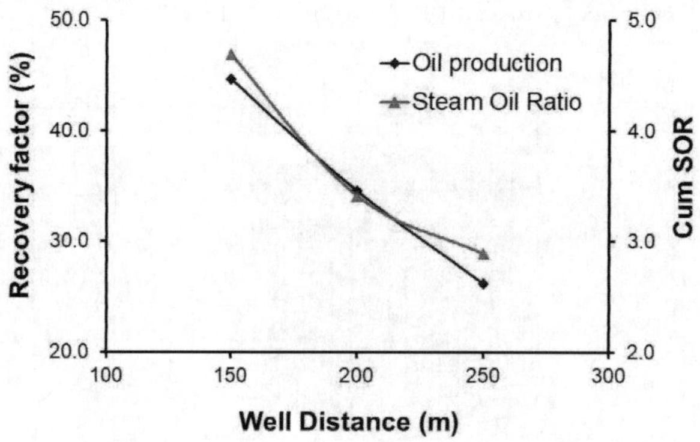

Figure 4. Effect of well distance on cumulated oil production and steam injection

It is evident that, compared to base case at 200 m, the cumulative oil production improves for 18 % when the area is larger and reduces 27 % when using smaller area. However, the cumulative steam oil ratio will increase approximately for 37 % for small area and decreases 15 % for 250 m well distance. Moreover, the well distance at 150 m consume high SOR. As a result, the cost of steam will be increased.

4.2.3 Effect of steam injection rate on oil production. Figure 5 illustrates the effects of oil recovery and the cumulative SOR for various steam injection rates ranging from 20-100 m^3/d. It is noted that the higher steam injection rate can increase both the total oil recovery and SOR. The injection rate at 100 m^3/d yields the highest RF at 43.58 %. However, the cumulative SOR is also the highest at 4.48 which means the cost of steam would be higher as well.

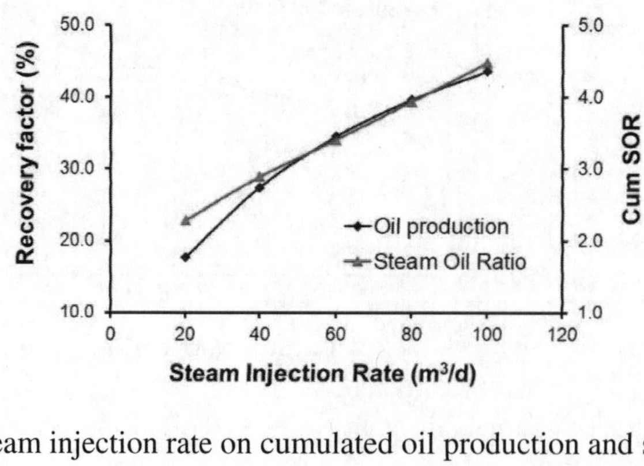

Figure 5. Effect of steam injection rate on cumulated oil production and steam injection

From the results, the cost of steam will be increased. Consequently, for practical work, to reduce the steam-oil ratio, the injection rate can be lowered with high permeability reservoirs. However, the economic point of view is required to decide the optimal condition for this area by consideration from the previous study with steam injection [11-12]. Subsequently, the conditions can be applied in the real field for heavy oil production.

5. Conclusions

Oil viscosity and its emulsions are measured and correlated over the entire temperatures ranging from 15 to 180°C. When steam is used, emulsion is formed and effect of hydrogen bonds on the viscosity has been explained. Moreover, the experimental data are correlated as a function of temperature based on the form in literature. The average absolute deviation is calculated at 12.92 %. Therefore, the correlation is suitable to represent the experimental data of viscosity of heavy oils and their emulsions.

For the simulation, the fluid properties and reservoir conditions are required to perform the simulation. Also, the parameters such as permeability, well distance and steam injection rates are essentially needed to be investigated for their effects on heavy oil production. The five-spot pattern is selected for simulation because it is effective and economical method for heavy oil production. From the simulation results, all parameters play a key role in heavy oil production. The recovery factor increases as permeability and injection increase and well distance decreases. Furthermore, steam-oil ratio would be reduced if injection rate is lowered with high permeability reservoirs. These conditions can be applied in the real field for heavy oil production. The main contribution of this practice is to reduce the steam consumption and fuel costs and to increase more oil production as well as to prolong the project life.

6. Acknowledgement

The authors would like to acknowledge Chevron Thailand Exploration and Production Ltd., for financial support of this project.

7. References

[1] Dusseault M B 2001 *The petroleum society's canadian international petroleum conference*, Calgary, Alberta, Canada, June 12-14

[2] Donaldson E C, Chilingarian G V and Yen T F 1989 *Enhanced oil recovery, ii processes and operations* (Amsterdam: Elsevier)

[3] Sheng J 2013 *Enhanced oil recovery field case studies* 1st ed. (Amsterdam: Gulf Professional Publishing)

[4] Ali S M and Meldau R F 1979 *Journal of Petroleum Technology* **31** 101

[5] Butler R M 1991 *Thermal recovery of oil and bitumen* (New Jersey: Prentice Hall)

[6] Garcia B, Alcalde R, Leal J M and Matos J S 1997 *J. Phys. Chem. B* **101** 7991

[7] Juntarasakul O and Maneeintr K 2018 *IOP Conf. Series: Earth and Environmental Science* **140** doi :10.1088/1755-1315/140/1/012024

[8] De Ghetto G, Paone F and Villa M 1995 SPE 30316, *The 1995 SPE international heavy oil symposium*, Calgary, Alberta, Canada, June 19-213

[9] Shen Bennison T 1998 *The IBC heavy oil field development conference*, London, December 2-4

[10] Lyons W C and Plisga G J 2005 *Standard handbook of petroleum & natural gas engineering vol. 2*, 2nd ed. (Houston, Texas: Gulf Publishing)

[11] Galvao E R V P, Rodrigues M A F, Dutra T V and da Mata W 2014 SPE-169442-MS *SPE Latin America and Caribbean Petroleum Engineering Conference*, 21-23 May, Maracaibo, Venezuela

[12] Chaar M, Venetos M, Dargin J and Palmer D B 2014 SPE-172004-MS *Abu Dhabi International Petroleum Exhibition and Conference*, 10-13 November, Abu Dhabi, UAE

IOP Conf. Series: Materials Science and Engineering **521** (2019) 012012 doi:10.1088/1757-899X/521/1/012012

Finite Element Analysis of Single Point Incremental Forming (SPIF) of Aluminum 7075 Using Different Types of Toolpath

R Esmaeilpour[1], F Pourboghrat[1,2], H Kim[1], T Park[1]

[1] Department of Integrated Systems Engineering, The Ohio State University, 210 Baker Systems, 1971 Neil Avenue, Columbus, OH 43210, United States
[2] Department of Mechanical and Aerospace Engineering, The Ohio State University, 201 W 19th Ave, Columbus, OH 43210, United States

Corresponding author e-mail: esmaeilpour.1@osu.edu

Abstract. To make large number of parts in automotive, aerospace, and appliances industries, sheet conventional metal forming processes, such as stamping, are used. However, for small number of productions, novel flexible manufacturing processes, such as incremental sheet forming (ISF) must be utilized to decrease the production costs. Studying the process parameters in ISF process such as, the toolpath, step size, tool size, feed rate, lubrication, and wall angle is essential as they are affecting the formability and deformation behavior of this process. The purpose of this study was to investigate the effect different types of toolpath, such as z-level (profile) and helical (spiral) to form a truncated-conical geometry in simulation of the single point incremental forming (SPIF) of AA7075-O. A detailed comparison of these tool paths' predictions was made with respect to the effective plastic strain distribution, part thickness, tool force and moment.

1. Introduction

In the last two decades, employing computers in sheet metal forming methods have led to flexible forming processes evolved from conventional ones, such as stamping. One of these novel methods is incremental sheet forming (ISF) process, in which the sheet metal is progressively being formed by CNC operated forming tool [1,2]. This method does not need any devoted punches or dies, [3] and brings the sheet suitable amount of plastic deformations which forms the final shape of the part. Due to the ISF method flexibility, producing broad 3D contours is feasible [4,5]. Single point incremental forming (SPIF) is one of the typical configurations of ISF, in which the sheet is formed with a single indenter [6].
Beside the experimental research, modeling of the different types of sheet forming processes, such as stamping and incremental forming process is vital to bring a better outlook of the deformation behavior and to optimize the forming conditions [7-11]. The purpose of this study was to investigate the effect of toolpath type as an important process parameter on single point incremental sheet forming (SPIF) process. For this reason, z-level (profile) and helical (spiral) tool paths were applied in simulation of the single point incremental forming (SPIF) to form a truncated-conical geometry from 7075-O aluminum alloy sheet. A detailed comparison of predictions was made with respect to the tool force and moment, effective plastic strain distribution and part thickness.

2. Uniaxial tensile tests

In previous studies by the authors, uniaxial tensile tests with standard ASTM rectangular dog-bone shaped samples of the aluminum alloy AA7075-O were performed to obtain the stress-strain curves and anisotropic properties at room temperature [12]. Tests were done with universal load frame INSTRON 5985. To monitor the specimens deformation, a 3D system digital image correlation (DIC), ARAMIS 5M, was used. The tests were repeated two times for each condition. All the tests were done at a strain rate of $0.01s^{-1}$. The uniaxial tensile tests engineering stress–strain curves for of AA7075-O in three different directions with respect to the rolling direction are shown in Figure 1.

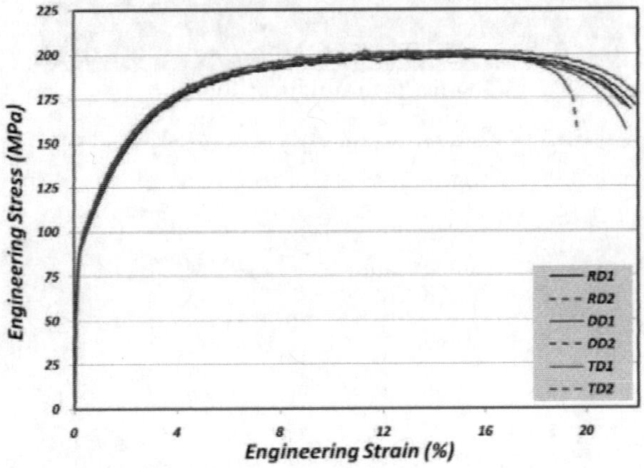

Figure 1. Engineering stress–strain curves in uniaxial tensile tests of the AA7075-O for RD, DD and TD [12]

3. Finite Element Simulation of SPIF Process

An explicit dynamic approach was used to simulate the SPIF process to produce a conical geometry with 67° angle, 55 mm depth, and 140 mm base diameter with a tool, modeled as a rigid body, with hemispherical head of 12.7 mm in diameter and feed rate of 50 mm/s. The sheet metal was assumed to be AA7075-O with the elastic properties of $E = 71.7\ GPa\ and\ \nu = 0.33$ and the density of 2.81E-9 ton/mm^3 [13]. The sheet is a circle with 90 mm radius, and a 1.6 mm thickness. Figure 2 represents the some shots of forming a conical part with SPIF process in Abaqus. For modeling the blank sheet, 87,773 C3D8R elements with dimensions of 0.7mm X 0.7mm X 0.5333mm (through-the-thickness) were used. All the elements at the blank boundaries were considered to be rigidly fixed to simulate the role of blank holder. The fixed mass scaling factor of 1.0E7 was applied. A coulomb friction coefficient of 0.1 was used, as commonly assumed in well lubricated sheet metal forming simulations.

It seems that because of through-the-thickness shears, considering a 3D yield function, like Yld2004-18p, is essential. However, as predicted strain histories were almost identical for various yield functions in previous studies by the authors, von Mises yield function was used in this study due to its simplicity and less computational time [12,14]. In addition, Voce type hardening law was used as it represented the material hardening behavior accurately. For this reason, the RD direction true stress-strain curve was used to calculate the parameters of the Voce type hardening law as shown here:

$$\bar{\sigma} = 94.5 + 136(1 - \exp(-24.5\bar{\varepsilon}) \quad (1)$$

where $\bar{\sigma}$ is in MPa.

IOP Conf. Series: Materials Science and Engineering **521** (2019) 012012 doi:10.1088/1757-899X/521/1/012012

Figure 2. Simulation shots

In ISF processes, designing the desirable tool paths are needed to provide optimal range of geometric accuracies in the formed part [15]. Two common types of toolpath i.e., spiral (helical) and z-level (profile) tool paths are shown in Figure 3.

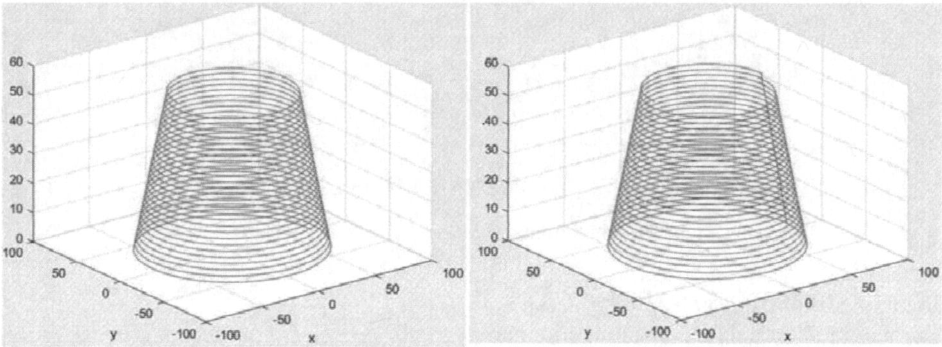

Toolpath Spiral **Toolpath Z-Level**

Figure 3. Common tool paths in the ISF process: left) Spiral (Helical), right) Z-Level (Profile)

These two types of toolpath were used in the FE simulation of SPIF process. A MATLAB code was written to generate these toolpaths to be used in Abaqus. Then, SPIF process was applied to form a conical geometry. Figure 4 shows effective strain distributions from the SPIF simulations of the sheet metal surface in contact with the tool for different toolpath types. As can be seen, the contour of the z-level toolpath presents a line of severe effective strain distribution which can make a scarring on the surface due to the upward transition between the following contours. Figures 5 and 6 present the tool reaction force and moment for different toolpath types, which are almost identical for spiral and z-level toopaths. As the excessive thinning of sheet metal makes failure in ISF process, the thickness was calculated as the distance from the inner to the outer sheet surface in direction of the normal to the surface, which is in contact with the tool. The determined thickness distribution for various toolpaths is presented in Figure 7. More thinning can be seen in the z-level toolpath thickness distribution, which shows this toopath has a higher failure chance. Before the deformation, the sheet thickness was 1.6 mm, while for the formed sheet, it is about 0.4 mm in some areas.

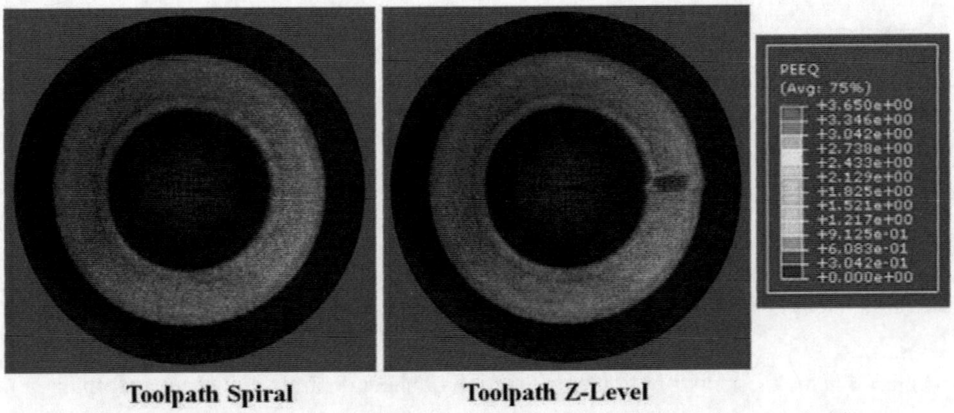

Figure 4. Sheet metal effective strain distribution using various toolpaths

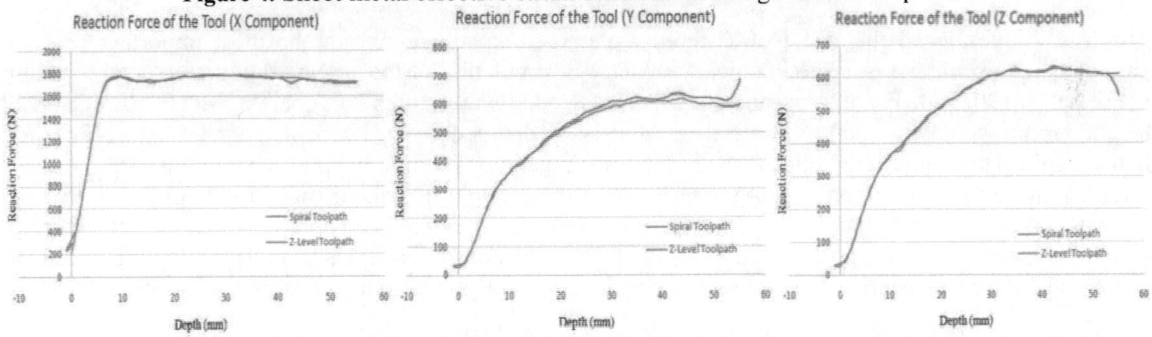

Figure 5. Components of the tool reaction force using various toolpaths

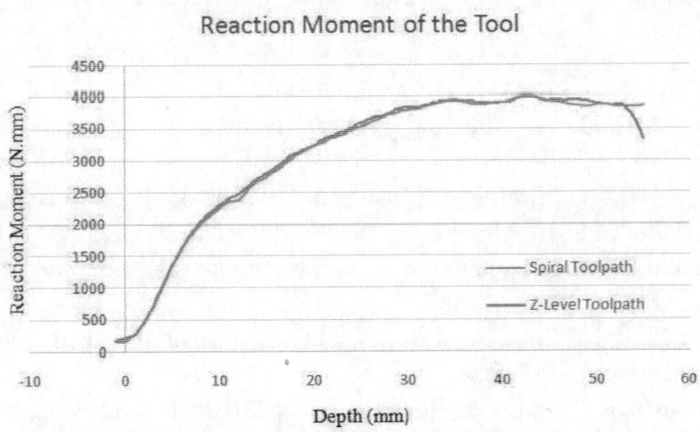

Figure 6. The tool reaction moment using various toolpaths

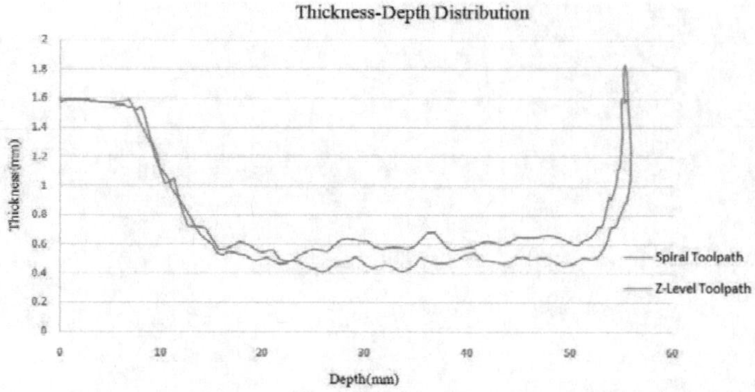

Figure 7. Thickness distribution as a function of sheet depth using various toolpaths

4. Conclusions

This study investigated the effect of different types of toolpath, as one of the most important process parameters, in forming a truncated-conical geometry in simulation of the single point incremental forming (SPIF) of AA7075-O. For this purpose, a MATLAB code was written to generate z-level (profile) and helical (spiral) toolpath types to be used in Abaqus. A detailed comparison of the z-level (profile) and helical (spiral) tool paths' predictions was made with respect to the effective plastic strain distribution, part thickness, tool force and moment. Results showed that due to the gradual transition between following contours, helical (spiral) tool path generates homogeneous thinning, and removes the surface scarring which was seen in z-level (profile) toolpath. It can be concluded that helical (spiral) tool path is a decent choice for ISF processes. However it might be difficult to be used to produce complex geometries.

Acknowledgement

The authors would like to acknowledge the financial support they received from the ONR ALMMII/LIFT funding for this research project.

References

[1] Adams D, Jeswiet J 2015 Design rules and applications of single-point incremental forming *Proc. Inst. Mech. Eng. Part B J. Eng. Manuf.* **229** 754–760

[2] Jeswiet J, Micari F, Hirt G, Bramley A, Duflou J, Allwood J 2005 Asymmetric Single Point Incremental Forming of Sheet Metal *CIRP Ann. - Manuf. Technol.* **54** 88–114

[3] Park J J, Kim Y H 2003 Fundamental studies on the incremental sheet metal forming technique *J. Mater. Process. Technol.* **140** 447–453

[4] Filice L, Fratini L, Micari F 2002 Analysis of material formability in incremental forming *CIRP Ann. - Manuf. Technol.* **51** 199–202

[5] Emmens W C, Sebastiani G, van den Boogaard A H 2010 The technology of Incremental Sheet Forming-A brief review of the history *J. Mater. Process. Technol.* **210** 981–997.

[6] Hagan E, Jeswiet J 2003 A review of conventional and modern single-point sheet metal forming methods *Proc. Inst. Mech. Eng. Part B J. Eng. Manuf.* **217** 213–225

[7] Kim H, Park T, Esmaeilpour R and Pourboghrat F 2018 Numerical Study of Incremental Sheet Forming Processes *J. Physics: Conf. Series.* **1063** 012017

[8] Pourboghrat F, Park T, Kim H, Mohammed B, Esmaeilpour R and Hector Jr L G, 2018 An integrated computational materials engineering approach for constitutive modelling of 3rd generation advanced high strength steels *J. Physics: Conf. Series.* **1063** 012010

[9] Park T, Kim H, Pourboghrat F and Esmaeilpour R 2018 A multi-scale modelling of 3rd generation advanced high strength steels to account for anisotropic evolution of yield surface and plastic potential *J. Physics: Conf. Series.* **1063** 012019

[10] Mohammed B, Park T, Kim H, Pourboghrat F and Esmaeilpour R 2018 The forming limit curve for multiphase advanced high strength steels based on crystal plasticity finite element modeling *Mater. Sci. Eng. A* **725** 250–66

[11] Mohammed B, Park T, Pourboghrat F, Hu J, Esmaeilpour R and Abu-farha F 2017 Multiscale crystal plasticity modeling of multiphase advanced high strength steel strength steel *Int. J. Solids and Structures.***000** 1–19

[12] Esmaeilpour R, Kim H, Park T, Pourboghrat F, Xu Z, Mohammed B and Abu-Farha F 2018 Calibration of Barlat Yld2004-18P Yield Function Using CPFEM and 3D RVE for the Simulation of Single Point Incremental Forming (SPIF) of 7075-O Aluminum Sheet *Int. J. Mech. Sci.* **145** 24–41

[13] http://asm.matweb.com/search/SpecificMaterial.asp?bassnum=MA7075O

[14] Esmaeilpour R, Kim H, Park T, Pourboghrat F and Mohammed B 2017 Comparison of 3D yield functions for finite element simulation of single point incremental forming (SPIF) of aluminum 7075 *Int. J. Mech. Sci.* **133** 544–54

[15] Blaga A, Bologa O, Oleksik V, Breaz R 2011 Influence of tool path on main strains, thickness reduction and forces in single point incremental forming process *Proc. Manuf. Syst.* **6** 2–7

IOP Conf. Series: Materials Science and Engineering **521** (2019) 012013 doi:10.1088/1757-899X/521/1/012013

Assembling Components with Probabilistic Lead Times

Noura Yassine[1]

Silvana El-Rabih[1]

[1] Beirut Arab University. P.O. Box 11-5020 Riad El Solh, Beirut 1107 2809, Lebanon

noura.yassein@bau.edu.lb

Abstract. This paper examines a production model in which *N* different types of components are used to assemble a finished product. The components are acquired from various suppliers in lots received at or before the beginning of the assembly process. The lead time between placing and receiving each of the *N* orders is assumed to be a random variable having a known probability density function. Based on recent results regarding the probability distribution and the expectation of the maximum of a set of independent random variables, the mathematical model describing this production/inventory situation is developed. The mathematical model is then used to derive a closed form formula for the optimal solution that minimizes the total production/inventory cost function. Moreover, the reorder point is shown to depend on the probability distribution and expected value of the random variable representing the maximum among the *N* lead times. The case in which each of the *N* lead times follows a Weibull distribution is investigated, and a numerical example is given to illustrate this case.

1. Introduction

The classical economic production quantity (EPQ) model determines the optimal production lot size that minimizes the total production/inventory cost by balancing the holding and setup costs. It is based on simplifying and unrealistic assumptions. By relaxing these assumptions, numerous research studies have been carried out so that the modified EPQ model accounts for factors faced in real life.

The classical EPQ model does not account for the various costs of components/raw material parts used in the production process. Salameh and El-Kassar [1] considered an EPQ that uses a single type of raw material. Salameh and Jaber [2] presented an inventory model in which each lot received from the supplier contains some imperfect items that can be salvaged at a discounted price. Khan et al. [3] presented a survey of articles that extend the model of Salameh and Jaber [2]. Several research studies have studied the effect of quality on the EPQ model (Salameh and Jaber [2]; Khan et al. [3]; Yassine et al. [4]). Most extensions of the EPQ model that account for quality of components/raw material parts considered a production process with a single type of components. Yassine [5] investigated a production process that uses different types of components and examined the effect of the quality of the components as well as the emissions tax incurred from transportation. Yassine [5] showed that the optimal production quantity depends on the expected value of the minimum of a set of random variables related to the percentages of perfect quality components of the various types.

This paper examines a production model that uses *N* different types of components in the assembly of the finished product. The components are acquired from various suppliers in lots received at or before the beginning of the assembly process. For each type of components, the lead time between placing and receiving the order is assumed to be a random variable having a known probability density function. A mathematical model describing this production/inventory situation is developed by applying results similar to those used in Yassine [6] and Yassine [5]. A closed form formula for the optimal solution is derived based on minimizing the total production/inventory cost function. Moreover, the reorder point is shown to depend on the probability distribution and expectation of the random variable representing the maximum among the lead times. The particular case where each of the lead times has a Weibull distribution is examined. Numerical examples are given to illustrate the calculation of the optimal solution and the reorder point.

2. Mathematical Model

In this section, the mathematical model is developed and the optimal order quantity is derived by minimizing the total inventory cost per unit time function.

2.1. Notation

The following notation will be used:

D Demand rate of the finished product (units/unit time)
P Assembly rate (units/unit time)
y Ordering quantity of the finished product (units)
y^* Optimal ordering quantity of the finished product (units)
C Cost of assembling one finished product ($/unit)
C_0 Set up cost of assembly ($)
C_i Purchasing cost per component of type i ($/unit)
K_i Order cost of components of type i ($)
h_i Holding cost of a component of type i per unit time ($/unit/unit time)
h Holding cost of one finished product per unit time ($/unit/unit time)
T Length of the inventory cycle (unit time)
T_p Length of the assembly period (unit time)
L_i Lead time of components of type i (unit time)
$f_i(L_i)$ Probability density function of L_i
μ_i Expected value of L_i (unit time)
σ_i Standard deviation of L_i (unit time)
L Lead time between placing and receiving all components (unit time)
μ_L Expected lead time of L (unit time)
R Reorder point (units)
$1-\alpha$ Service level
SS Safety stock of finished items (units)

2.2. Assumptions

The development of the mathematical model is based on the following assumptions:

- The finished product is assembled using N types of components.
- A single item of the finished product requires one unit of each type of components.
- All components received from the suppliers contain some imperfect quality items.
- The finished product demand rate D is deterministic.
- Assembly time is significant.
- The lead time L_i for each type of components is a random variable having a known probability distribution $f_i(L_i)$.
- An order of size y of each type of components is placed whenever the inventory level of the finished reaches a reorder point R.

2.3. The Total Cost per Unit Time Function

The objective of this model is to determine the optimal quantity y^* of the finished product as well as the reorder point R. The lead time L is defined to be the time between placing and receiving the N orders is given by

$$L = \max(L_1, L_2, ..., L_N). \tag{1}$$

IOP Conf. Series: Materials Science and Engineering **521** (2019) 012013 doi:10.1088/1757-899X/521/1/012013

Given a required service level of $(1\text{-}\alpha)\times 100\%$, the reorder point R is determined by

$$P(LD > R) = \alpha. \tag{2}$$

This reorder point R is the sum of the expected demand during lead time, $\mu_L D$, and the safety stock, SS, that guaranties the required service level of $(1\text{-}\alpha)\times 100\%$. That is,

$$R = \mu_L D + SS. \tag{3}$$

The total cost per inventory cycle function, $TC(y)$, is the sum of the setup cost of assembly, the costs of ordering the components, as well as the holding costs of components, finished product, and the safety stock. From Figure 1, the $TC(y)$ function is given by

$$TC(y) = C_0 + Cy + \sum_{i=1}^{N}(C_i y + K_i) + \frac{h}{2}(P - D)T_p T + h(R - \mu_L D)T + \sum_{i=1}^{N} h_i\left((L - L_i)y + \frac{y}{2}T_p\right). \tag{4}$$

The total cost per unit time function, $TCU(y)$, is obtained by dividing the expression in equation (4) by the inventory cycle length T. The expected total cost per unit time function $ETCU(y) = E(TC(y)/T)$. By the Renewal Reward Theorem, $ETCU(y)$ is approximated by $E(TC(y))/E(T)$. Since the order quantity is y, the expected cycle length is y/D and the expected production period is y/P. Hence, $E(T) = y/D$, $E(T_p) = y/P$, and the expected total cost per unit time function $ETCU(y)$ is

$$ETCU(y) = \left(\frac{C_0}{y} + C\right)D + \sum_{i=1}^{N} D\left(C_i + \frac{K_i}{y}\right) + \frac{yh}{2}\left(1 - \frac{D}{P}\right) + h(R - \mu_L D) + \sum_{i=1}^{N} Dh_i\left((\mu_L - \mu_i) + \frac{y}{2P}\right). \tag{5}$$

Differentiating $ETCU(y)$,

$$\frac{d}{dy}(ETCU(y)) = -\frac{C_0 D}{y^2} - \sum_{i=1}^{N}\frac{K_i D}{y^2} + \frac{h}{2}\left(1 - \frac{D}{P}\right) + \sum_{i=1}^{N} h_i \frac{D}{2P}. \tag{6}$$

Setting the derivative in equation (6) equal to zero and solving for y, the optimal order quantity of the finished product y^* is

$$y^* = \left(\frac{C_0 D + \sum_{i=1}^{N} K_i D}{\frac{h}{2}\left(1 - \frac{D}{P}\right) + \sum_{i=1}^{N} h_i \frac{D}{2P}}\right)^{1/2}. \tag{7}$$

Note that the second derivative of $ETCU(y)$, obtained by differentiating the equation (6), is always negative. Hence, y^* is the unique minimizer of $ETCU(y)$.

3. Lead Times with Weibull Distributions

Among the probability distributions identified in the literature for the lead time is the Weibull distribution [7].

Let L_1 and L_2 be two independent random variables having Weibull distributions with positive scale parameters α_1 and α_2 and identical positive shape parameters each equal to k. It can be shown that the probability density function of $L = max(L_1, L_2)$ is

$$f_L(l) = \frac{k}{\alpha_1}\left(\frac{l}{\alpha_1}\right)^{k-1}.\exp\left(-\left(\frac{l}{\alpha_1}\right)^k\right) + \frac{k}{\alpha_2}\left(\frac{l}{\alpha_2}\right)^{k-1}.\exp\left(-\left(\frac{l}{\alpha_2}\right)^k\right) - \frac{k}{\alpha^*}\left(\frac{l}{\alpha^*}\right)^{k-1}.\exp\left(-\left(\frac{l}{\alpha^*}\right)^k\right) \tag{8}$$

IOP Conf. Series: Materials Science and Engineering **521** (2019) 012013 doi:10.1088/1757-899X/521/1/012013

where $x \geq 0$ and $\alpha^* = \left(\frac{1}{\alpha_1^k} + \frac{1}{\alpha_2^k} \right)^{-\frac{1}{k}}$. Moreover, cumulative distribution of L is

$$F_L(t) = 1 - .\exp\left(-\left(\frac{l}{\alpha_1} \right)^k \right) - \exp\left(-\left(\frac{l}{\alpha_2} \right)^k \right) + \exp\left(-\left(\frac{1}{\alpha_1^k} + \frac{1}{\alpha_2^k} \right) l^k \right), \tag{9}$$

and the expected value of L is

$$\mu_L = (\alpha_1 + \alpha_2 - \alpha^*)\Gamma\left(1 + \frac{1}{k} \right)^{k-1}. \tag{10}$$

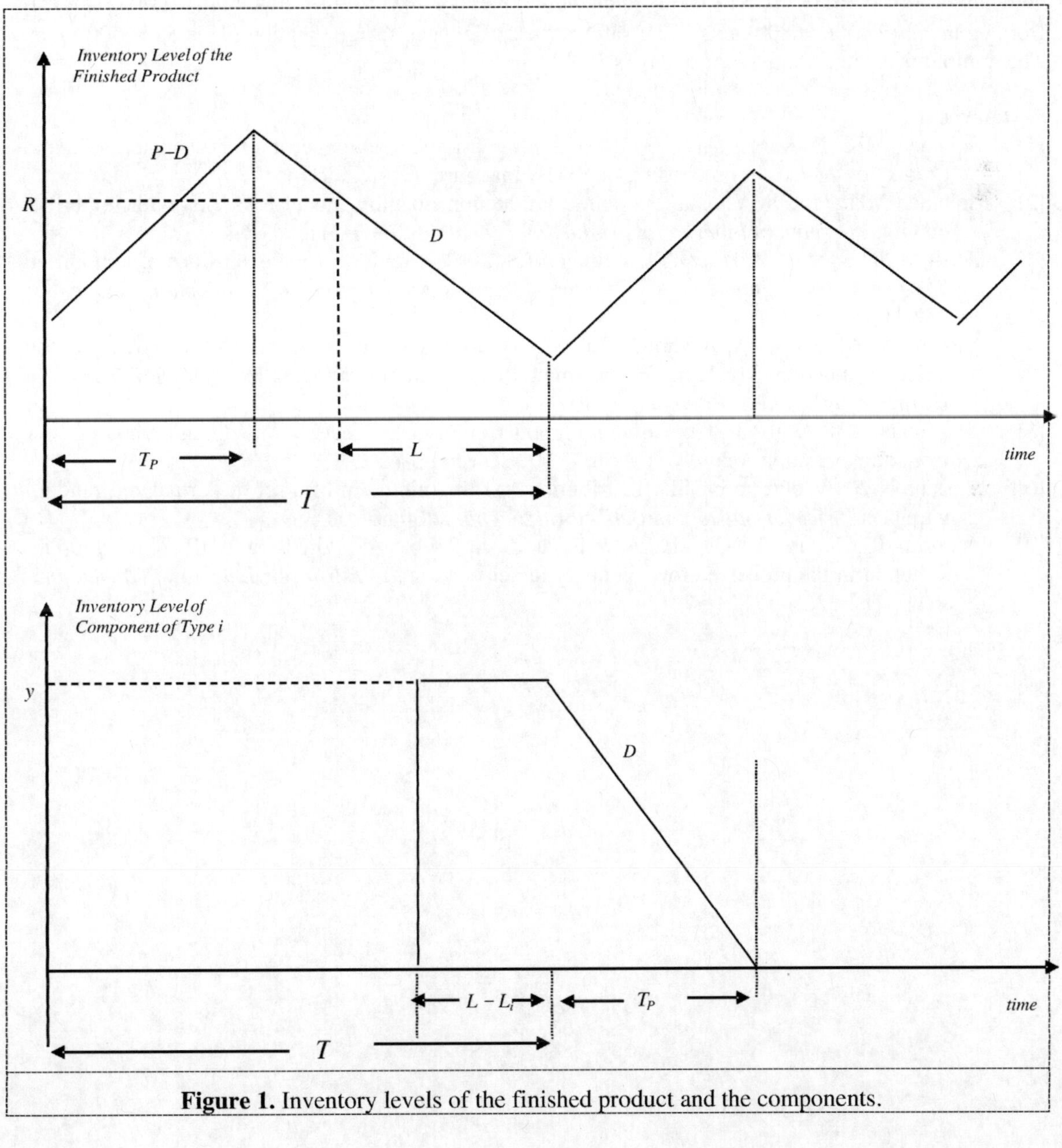

Figure 1. Inventory levels of the finished product and the components.

4. Numerical Example

Suppose that the demand rate of the finished product is 200 units per day. A single item of each of the two types of components, 1 and 2, are assembled to form a finished item. The costs of ordering components of type 1 and type 2 are \$200 and \$300, respectively. The unit purchasing costs of components of type 1 and type 2 are \$2 and \$3, respectively. The holding cost per unit per day is \$0.01 for component 1, \$0.02 for component 2, and \$0.06 for the finished product. The assembly cost is \$4 per item of the finished product. The assembly rate is 400 units/day. The optimal order quantity obtained from equation (6) is $y^*= 3{,}651$ units.

Suppose that the lead times in months follow Weibull distributions with parameters $\alpha_1 = \alpha_2 = k = 1$. Then, the expected lead times μ_1 and μ_2 are both equal to 1 month. From equation (9), the expected value of L is $\mu_L = 1.499$ months. Substituting the parameters of the distributions in equation (8), the cumulative distribution of L is $F_L(t) = 1 - 3e^{-t}$. For a service level of 95%, the reorder point obtained from equation (2) is $R = 800$ units. The safety stock calculated using equation (3) is $SS = 500$ units. The minimum expected total cost per day is \$1,998.40.

5. References

[1] Salameh MK, El-Kassar AN 2007 Accounting for the holding cost of raw material in the production model In Proceeding of BIMA Inaugural Conference pp 72-81

[2] Salameh MK, Jaber MY 2011 Economic production quantity model for items with imperfect quality. *International journal of production economics.* **64(1-3)** pp 59-64

[3] Khan M, Jaber MY, Guiffrida AL, Zolfaghari S. 2011 A review of the extensions of a modified EOQ model for imperfect quality items *International Journal of Production Economics* **132(1)** pp 1-2

[4] Yassine N, AlSagheer A, Azzam N 2018 A bundling strategy for items with different quality based on functions involving the minimum of two random variables *International Journal of Engineering Business Management* **10** pp 1-9

[5] Yassine N 2018 A sustainable economic production model: effects of quality and emissions tax from transportation *Annals of Operations Research* pp 1-22.

[6] Yassine N 2016 Joint probability distribution and the minimum of a set of normalized random variables *Procedia-Social and Behavioral Sciences* **230** pp 235-239

[7] Amorim P, Curcio E, Almada-Lobo B, Barbosa-Póvoa AP, Grossmann IE 2016 Supplier selection in the processed food industry under uncertainty *European Journal of Operational Research* **252(3)** pp 801-814

IOP Conf. Series: Materials Science and Engineering **521** (2019) 012014 doi:10.1088/1757-899X/521/1/012014

Multi-Objective Scheduling Simulation of Adaptive Job Shop based on Modified SOMA Algorithm

Danding Jiang[1], Mingwei Wang[2], Ying Zhao[3] and Tengyuan Jiang[2]

[1] College of Systems Engineering, National University of Defense Technology, Changsha, China

[2] School of Mechanical Engineering, Northwestern Polytechnical University, Xi'an, China

[3] China Aerospace Electronic Technology Research Institute, Beijing, China

E-mail:jddname@aliyun.com

Abstract. In this paper, an effective modified Self-Organization Migrating Algorithm (SOMA) is proposed to solve multi-objective adaptive job shop scheduling with the criterion to minimize the processing cost, minimize makespan, minimize the total machine loads. The modified SOMA stresses the balance between global exploration and local exploitation by introduce adaptive step, and improved the population diversity in the process of individual migration by introduce quadratic interpolation, which effectively avoided the premature and improved the convergence of the SOMA. Finally, through the 4×6 job shop scheduling problem verify the performance of the modified SOMA algorithm, the computational results show that the proposed modified SOMA efficiently solves adaptive job shop scheduling.

1. Introduction

With the change of production model to small batch, multiple varieties and customization, the job shop is required to be able to adjust quickly according to the market demand in order to adapt to the fast tempo production mode, therefore, multi-objective adaptive job shop scheduling has attracted a lot of attention [1-7]. For instance, Ahmed Farouk etc. provide a nonlinear GA optimization model for allocation of resources and available time to tasks for scheduling project [1]. Imen Driss etc. propose a new genetic algorithm to solve FJSP to minimize makespan [2]. Miguel A etc. propose a new hierarchical heuristic algorithm which is an adaptation of the Newton's method for continuous multi-objective unconstained optimization problems for multi-objective FJSP [3]. Liu A J etc. established mathematical model according to the objective function of minimizing the tardiness penalty and manufacturing time, employed the hybrid and cycle driven rescheduling to modified genetic algorithm optimized the process route and processing sequences[4]. Piroozfard H, etc. used the modified multi-objective genetic algorithm to complete the job shop scheduling problem which use carbon emissions and total delay time as scheduling objective [5]. C Zhu etc. for optimized the scheduling stability and robustness of job shop, used the multi objective differential evolution algorithm to solve the scheduling of flexible job shop [6]. Hadi Mokhtari etc. aimed the target of makespan, system utilization and total energy cost, completed the scheduling problem by combined with enhanced evolutionary algorithms and global standards [7]. In conclusion, for balance multiple objectives to obtain the optimization, the multi objective need to transformed into single objective scheduling problem. Under the guidance of this method, overall consideration minimize the processing cost, minimize makespan, minimize the total machine loads, realization the self-adaptation of scheduling process. After the SOMA has been proposed to optimization problem, it widely used in many fields due to its strengths such as fast convergence speed, few adjustment parameters, and the ability of self-adaption [8]. Davendra D etc. have used the SOMA to solve the flow-shop scheduling with no-wait makespan [9]. The modified SOMA stresses the balance between global exploration and local exploitation by introduce adaptive step; improved the population diversity in the process of individual migration by introduce quadratic interpolation. Finally, the computational results show that the proposed modified SOMA efficiently solves job shop scheduling.

2. Adaptive Job Shop Scheduling

2.1. Job Shop Scheduling Description.

The job shop scheduling problem can be described as following: the jobs set $J=\{J_1,J_2,\cdots,J_n\}$, where n is the total number of workshop jobs; the machine set $M=\{M_1,M_2,\cdots,M_m\}$, where m is the total number of machine in the workshop; processes set $po_i=\{po_{i1},po_{i2},\cdots,po_{in_b}\}$ is the processes sequence of the job J_i, where n_{bi} represents the number of processes in the job, $po^k_{i,j}$ represents the j-th process of job J_i is processed on machine k.

2.2. Multi-Objective Optimization Model

Most of the literature use the makespan, cost as indicators to evaluate the scheduling scheme. Although these indicators can reflect the merits of the scheduling scheme at a large extent, but leave the machine load out of consideration. This paper analyzes the existing literature and considers the three goals of production cost, makespan and total machine loads to express the cost, time, and machine load of job shop respectively. The objective function is as follows:

- Minimization production cost

$$Cost = Cost_{ready} + Cost_{process} = \sum_{k=1}^{m}\sum_{i=1}^{n}\sum_{j=1}^{n_{bi}}(cr^k_{i,j}\cdot po^k_{i,j}\cdot rt^k_{i,j} + cp^k_{i,j}\cdot po^k_{i,j}\cdot ot^k_{i,j}) \quad (1)$$

Where $Cost$ indicate the total cost of the job shop; $Cost_{ready}$ indicate the preparation cost; $Cost_{process}$ indicate the processing costs; $cr^k_{i,j}$, $cp^k_{i,j}$ are respectively indicates the preparation cost coefficient, the process cost coefficient of the j-th process of the job J_i on the machine k; $D^k_{i,j}$ indicates the deadline of the j-th process of the job J_i on the machine k.

- Minimize the makespan

$$T = \sum_{i=1}^{n}\sum_{j=1}^{n_b}\sum_{k=1}^{m}(rt^k_{i,j} + ot^k_{i,j}) \quad (2)$$

Where T represents the total processing time of the job shop, $rt^k_{i,j}$ represents the preparation time of the j-th process of the job J_i on the machine k, $ot^k_{i,j}$ represents the processing time of the j-th process of the job J_i on the machine k.

- Minimize total machine loads

$$\min W = \min \sum_{k=1}^{m} W_k \quad (3)$$

Where W is total machine loads, W_k is the load of machine k, all of load is measured by machining time, which W_k can be description as follow:

$$W_k = \sum_{i=1}^{n}\sum_{j=1}^{n_{bi}} po_{i,j}^k \cdot t_{i,j}^k \qquad (4)$$

Where $t_{i,j}^k$ is machining time of the j-th process of the job J_i on the machine k, which consist of preparation time and processing time.

Subject to:

- ✧ Processing time is not negative
- ✧ the preparation time is before the processing time
- ✧ The processing of the next process can only be carried after the completion of the previous process
- ✧ A machine tool cannot perform two job simultaneously
- ✧ The job must to be completed before the deadline

In order to achieve the balance among the various optimization objectives and obtain an approximate optimal scheme to the production process of the optimization process of job shop scheduling, therefore, integrated the above three optimization objectives form a single objective. Before integrating optimization objectives, needs to use the following formula to normalize the optimization objective:

$$a = \frac{b_{max} - b}{b_{max} - b_{min}} \qquad (5)$$

Where a is the normalized result of the optimization objective, b_{max} represents the maximum value of the optimization objective, b_{min} represents the minimum value of the optimization goal, b represents the current value of the optimization function.

Based on the above three normalized optimization goals, weights are used to normalize them to a single scheduling objective, as follows:

$$f = \omega_1 \cdot \overline{Cost} + \omega_2 \cdot \overline{T} + \omega_3 \cdot \overline{W_k} \qquad (6)$$

Where f represents the optimization objective, $\overline{Cost}$ represents the normalized cost, $\overline{T}$ represents the normalized makespan, $\overline{W_k}$ represents the total machine loads, $\omega_1, \omega_2, \omega_3$ are weight factors, which are obtained through an adaptive scheduling strategy and expert experience, $\omega_1 + \omega_2 + \omega_3 = 1$, $\omega_1 \geq 0, \omega_2 \geq 0, \omega_3 \geq 0$.

3. Modified SOMA-Based Scheduling Algorithm

In order to modify the self-adaptive ability of SOMA algorithm, this paper combines adaptive step size with quadratic interpolation to improve the exploration and diversity of SOMA respectively.

3.1 *Adaptive Step Setting*

In the individual migration process of SOMA, step represents the size of the search range of the individual in the optimal scheme direction, and in the standard SOMA algorithm, the recommended value is 0.11. However, during the actual simulation application process, it can be found that in the early stage of the algorithm operation, a larger step size can be used to improve the exploration capability of the algorithm. In the later stage of operation, a smaller step length is used to enhance the local search ability of the algorithm, and the local search ability of the algorithm is effectively utilized while ensuring the improvement of the algorithm's convergence speed. Therefore, this paper adopts the linear decreasing step strategy proposed in literature [10], so that the initial step length is the maximum step size $Step_{max}$, the termination step size is the minimum step size $Step_{min}$, and the step size is adaptively updated after each evaluation. The adaptive step length is defined as:

IOP Conf. Series: Materials Science and Engineering **521** (2019) 012014 doi:10.1088/1757-899X/521/1/012014

$$Step = Step_{\max} - (\frac{Count}{Sum\_Count}) \cdot (Step_{\max} - Step_{\min}) \qquad (7)$$

Where $Count$ represents the current time of evaluation, Sum_Count represents the maximum time of evaluation, and according to the experimental results in [10], $Step_{\max}$ is set to 0.7 and $Step_{\min}$ is set to 0.4.

3.2 *Adaptive Individual Migration*

Different from others evolutionary algorithms, the SOMA algorithm use a real number PRT between [0, 1] as a perturbation to determine whether an individual migrate to a leader. The detailed introduction of PRT refer to the literature [8].

In order to improve the diversity of the population, refer to the literature [11] to introduce quadratic interpolation. When the perturbation matrix A is 0, the best individual (leader) and two random individuals in the population are selected, use the quadratic interpolation method generate a new individual. The formula is as follows:

$$x' = \frac{[(R_2^2 - R_3^2) \cdot f(R_1) + (R_3^2 - R_1^2) \cdot f(R_2) + (R_1^2 - R_2^2) \cdot f(R_3)]}{2[(R_2 - R_3) \cdot f(v_1) + (R_3 - R_1) \cdot f(R_2) + (R_1 - R_2) \cdot f(R_3)]} \qquad (8)$$

Where R_1 represents the leader, R_2 and R_3 represent individuals randomly selected from the population. If the new individual is accepted when it is better than the original individual and is replaced with original individual.

4. Analysis of Example

A job shop is composed of 6 machines and it needs to produce 4 kinds of jobs. Every job need three processes, and each process can select and process on three different machine. The deadline of jobs are shown in table 1, the preparation and processing cost coefficient are shown in table 2, the preparation and processing time of every process are shown in table 3.

Table 1. The deadline (hour) of jobs.

job	1	2	3	4
Deadline	19	15	19	20

Table 2. The cost coefficient of preparation and processing ($).

machine	1	2	3	4	5	6
Preparation	15	16	18	16	15	17
Processing	40	45	50	40	47	43

Table 3. The preparation and processing time (hour) of jobs.

job	process	Machine					
		1	2	3	4	5	6
	1	2,3	---	2,5	---	1,2	---
1	2	2,4	---	---	---	3,6	2,3
	3	---	---	1,4	---	2,7	3,4
	1	2,2	2,3	2,3	---	---	---
2	2	----	1,3	---	2,2	3,2	---
	3	1,1	2,4	3,5	---	---	---
	1	1,6	---	2,5	2,4	---	---
3	2	---	2,6	---	3,4	---	3,5
	3	2,2	---	2,3	---	---	3,3
	1	2,6	2,5	---	3,7	---	---
4	2	---	2,4	1,3	---	2,5	---
	3	---	2,5	1,6	---	2,7	---

The basic parameters of this modified SOMA algorithm are as follows: Stepmax=0.7, Stepmin=0.4, PRT=0.1, PathLength=3. Through the simulation of the algorithm, the optimal value of the objective function is 0.9609, the cost is 1991 and the completion time of jobs are 15,11,17,18h, the total machine load is 61.The Gantt chart of its scheduling is shown in figure 1.

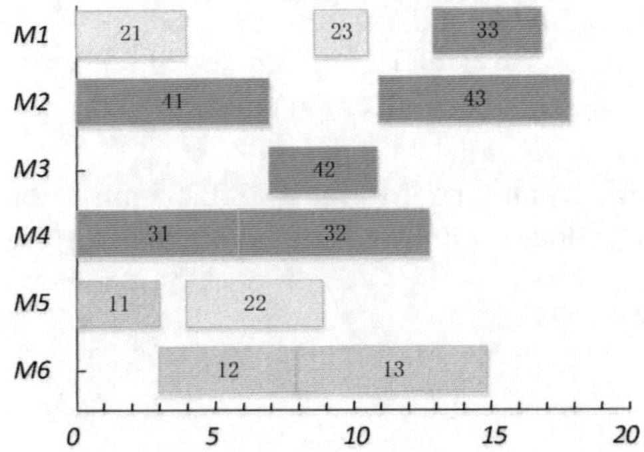

Figure 1. The Gantt chart of 4*6 scheduling problem.

5. Conclusion

This paper put forward an efficient modified SOMA algorithm based on adaptive step and quadratic interpolation, for solving multi-objective job shop scheduling problem with the criterion to minimize the processing cost, minimize makespan, minimize the total machine loads. The performance of the algorithm is improved effectively by adaptive step and quadratic interpolation, which use the adaptive step improve the ability of the exploration and exploitation, use the quadratic interpolation to improve the diversity of population. Finally, check this algorithm with example of 4×6 job shop scheduling, and the computational results show that the proposed modified SOMA efficiently solves job shop scheduling. In the future work will use the modified SOMA to solve the large scale adaptive job shop scheduling and try to combine with other algorithms to improve the ability to solve problem .

6. Acknowledgments

This work was supported by The Marine Diesel Engine Cylinder Head Precision Processing Flexible Production Line Development and Application Project(Number: 2012ZX04011-041).

7. References

[1] Driss I, Mouss K N, Laggoun A. A new Genetic Algorithm for Flexible Job-Shop Scheduling Problems. Journal of Mechanical Science & Technology, 2015, 29(3):1273-1281.

[2] Pérez M A F, Raupp F M P. A Newton-based Heuristic Algorithm for Multi-objective Flexible Job-Shop Scheduling Problem. Journal of Intelligent Manufacturing, 2016, 27(2):409-416.

[3] Ahmed Farouk, Khaled EI Kilany, Fady Safwat. Genetic Algorithm for Project Scheduling and Resource Allocation under Uncertainty. International journal of Mechanical Engineering and Robotics Research,Vol.2,No.3,July 2013

[4] Liu A J, Yang Y, Xing Q S, et al. Dynamic Scheduling on Multi-objective Flexible Job Shop. Computer Integrated Manufacturing Systems, 2011, 17(12):2629-2637.

[5] Piroozfard H, Wong K Y, Wong W P. Minimizing Total Carbon Footprint and Total Late Work Criterion in Flexible Job Shop Scheduling by Using an Improved Multi-Objective Genetic Algorithm. Resources Conservation & Recycling, 2016.

[6] C Zhu, W Qiu, C Zhang,et al. Multi-objective Flexible Job Shop Dynamic Scheduling Strategy aiming at Scheduling stability and robustness. China Mechanical Engineering, 2017.28(2):173-182.

[7] Mokhtari H, Hasani A. An Energy-Efficient Multi-Objective Optimization for Flexible Job-Shop Scheduling Problem. Computers & Chemical Engineering, 2017.

IOP Conf. Series: Materials Science and Engineering **521** (2019) 012014 doi:10.1088/1757-899X/521/1/012014

[8] Zelinka I. SOMA — Self-Organizing Migrating Algorithm, Self-Organizing Migrating Algorithm. Springer International Publishing, 2016:167-217.
[9] Davendra D, Zelinka I, Bialic-Davendra M, et al. Discrete Self-Organising Migrating Algorithm for flow-shop scheduling with no-wait makespan. American Institute of Physics Conference Series. American Institute of Physics, 2011:285-289.
[10] WENG Chunyi, LI Yuanxiang, WANG Lingling, et al. Modified self-organizing migrating algorithm with linear-digress step. Computer Engineering and Applications , 2011, 47（18）:26-28.
[11] Singh D, Agrawal S. Self-Organizing Migrating Algorithm with Quadratic Interpolation for Solving Large Scale Global Optimization Problems. Applied Soft Computing, 2016, 38:1040-1048.

Path Planning for Spray Painting Robot of Horns Surfaces in Ship Manufacturing

Yunzhong Zhou[a] *, Shumei[b] Ma, Aiping[c] Li and Liansheng[d] Yang

College of Mechanical Engineering, Tongji University, No.4800 Caoan Road, Jiading District, Shanghai, China

[a]zhouyunzhong@tongji.edu.cn, [b]mashumei@tongji.edu.cn, [c]limuzi@tongji.edu.cn, [d]shenge_yang@aliyun.com

Abstract. Generating paint gun path for complex free-form surfaces in shipbuilding industry to ensure uniform paint deposition is highly challenging due to their complex geometry, especially for horns surfaces. In this paper, a tool trajectory optimization method was proposed based on dip angle spray in order to solve the problems of poor coating uniformity and material waste for spraying horns surface with curved patches. A dip angle spray model was developed based on the Beta distribution model on the planar surface by principle of differential geometry. The spray path optimization models for dip angle spraying on cylindrical surfaces of three cases were proposed and compared first to define the best case which yields the best painting quality. Then we used the best strategy on a horns surface in a shipyard to evaluate the effectiveness and robustness of designed algorithm. This algorithm can also be extended to other applications.

1. Introduction

Spray painting robots are commonly adopted in many industries such as automobiles, furniture and shipbuilding industry to improve productivity as well as painting quality. However, the traditional "artificial teaching" style of spraying robots have largely limited their usage since it involves tedious teaching process which can be time-consuming and energy-inefficient, thereby making the off-line programming a hot topic among researchers [1].

The shape of workpiece and the tool parameters can strongly influence the quality of painting [2]. Some early researches studied the painting quality of plane of near-plane surfaces [3, 4, 5]. Later, Balkan [6] used experimental paint thickness distribution resulting from a single painting stroke applied on a flat surface to predict the painting thickness and found that Beta model was more accurate than traditional parabolic model. Zaki [7] developed an economic automatic system for spray painting of a general three dimensional surface. In order to improve the painting quality on free-form surfaces, Chen [8] proposed to split a free-form surface into several patches and conducted an experiment to verify the proposed method. However, the method was only applicable to simple free-form surfaces. Sheng [9] proposed a tool-path planning approach which optimizes the tool motion performance and the thickness uniformity of geometry-complicated parts consisting of multiple freeform surfaces. Chen [10, 11] pointed out that a complex free-form surface is consisted of several patches and then proposed trajectory integration algorithm which helped better the material quantity deviation. However, like other former researchers, he split a free-form surface into several near-plane patches which might cause low painting quality and painting efficiency problem especially when there are too many patches

IOP Conf. Series: Materials Science and Engineering **521** (2019) 012015 doi:10.1088/1757-899X/521/1/012015

due to the complex geometric character of a free-form surface. To solve the above-mentioned problem, Gong [12] developed an algorithm to divide complex free-form surfaces into several regular surfaces and established static models for regular surfaces. In order to improve the accuracy of the spray gun model, Zhang [13] developed a natural quadric mathematical model of arc spray gun trajectory considering gun spraying flow.Some scholars also proposed trajectory planning schemes with Bézier curves, which can also ensure painting quality for free-form surfaces [14,15]. However, the modelinng of Bézier curves is very time-consuming and its mathematical expression can be too complicated to generate satisfactory trajectories [16]. Fu [17] proposed a genetic algorithm-based surface segmentation method in which complex surface are divided into several small curvature surfaces and it simplified the subsequent trajectory planning operations. Zhou [18] discussed the automatically generated trajectories on regular surfaces using cutting-projection processing, visual simulation results demonstrate the effectiveness and practicality of the proposed algorithm. In order to solve the problems of poor coating quality and paint waste for spraying inner/outer horns surface, a tool trajectory optimization method was proposed based on dip angle spray. The simulation results illustrated the feasibility and availability of these optimization model and algorithms [19, 20]. However, they only assumed that the horns surface were composed of two near-plane patches. If one of the patches was actually a curvature surface, then their methods might not work out well. In order to solve the problems of coating growth rate modeling for varied dip-angle spraying technology, Zeng [21] developed a prediction mode of coating growth rate. Experiments have shown that the prediction model has good fitting precision.

In this paper, we proposed an algorithm to solve the problem of poor coating for spraying horns surfaces with curvature patches. Curvature patches can be fitted into cylindrical surfaces to simplify the problem without losing too much accuracy compared with being fitted into planes. The dip angle spray model for cylinders were constructed by principle of differential geometry. Based on the dip angle spray model, we then established a trajectory generation model and optimized the parameters to get the best painting quality. Simulation results showed that the proposed method yields satisfactory results.

2. Dip Angle Spray Modelling On Cylindrical Surfaces

Paint distribution model is needed in order to generate paint gun path. In view of the accuracy and flexibility, we consider a Beta distribution model. The paint flux q is given by equation (1) as follows:

$$q(r) = q_{\max}(1-\frac{4r^2}{w^2})^{\beta-1}, r \in \left[0, \frac{w}{2}\right]. \tag{1}$$

figure 1 shows paint accumulation of dip angle spray on a cylindrical surface. P_1 is a reference plane and P_2 is a parallel plane which passes the point B; α is the angle of gun axis and the line of the point B to gun center; h_s is the actual tool height; h is the desired tool height. β is the angle of gun axis and the normal vector of C, which is the dip angle of the spray gun. Suppose the material sprayed on a small area C_1 is projected to the area C_2. The film thickness of C_2 can be modeled as:

$$q_2 = q_1(S_1 / S_2) = q_1(\frac{h}{hs})^2. \tag{2}$$

figure 2(a) shows a circle C_3, which is a perpendicular to the material emission direction. The material on C_3 is projected to the free-from surface with deviation angle δ, as shown in figure 2(b). The material thickness on C_3 and C_4 can be expressed as:

$$q_3 = q_2 / \cos(\alpha); \quad q_4 = q_3 \cos(\delta). \tag{3}$$

Based on Equation (2), (3), the material thickness on the cylindrical surface can be obtained as:

$$q(r) = q_{\max}(1-\frac{r_s^2}{R^2})^{\beta-1}(\frac{h}{h_s})^2 \cos(\delta) / \cos(\alpha). \tag{4}$$

Where:

IOP Conf. Series: Materials Science and Engineering **521** (2019) 012015 doi:10.1088/1757-899X/521/1/012015

$$h_S = h + x\sin\beta \pm (\rho - \sqrt{\rho^2 - x^2}); \cos\beta = h + x\sin\beta \pm (\rho - z)\cos\beta; \quad \delta = \alpha \pm \gamma - \beta$$

$$r_s = \frac{h\sqrt{(x\cos\beta)^2 + y^2 \pm (z\sin\beta)^2}}{h_S} \tag{5}$$

$$\alpha = \arccos\left(\frac{h_S}{\sqrt{(x + h\sin\beta)^2 + y^2 + (\sqrt{\rho^2 - x^2} - (\rho + h\cos\beta))^2}}\right)$$

When the surface is a plane, ρ is infinite, and the dip angle spray model for planes can be obtained in Equation (4).

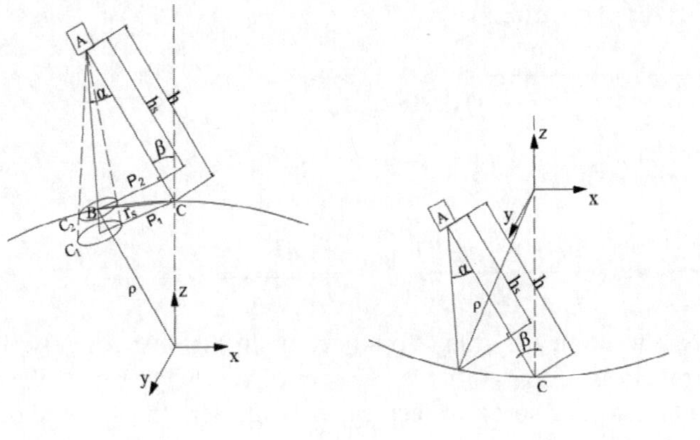

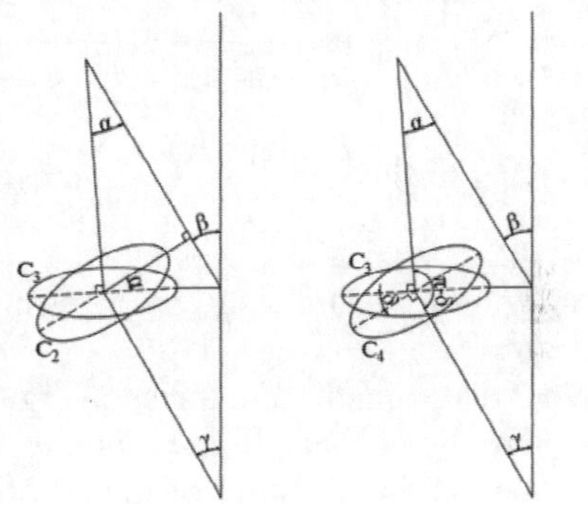

Figure 1. Profile of dip angle spray process on cylindrical surfaces. **Figure 2.** Material projection.

3. Optimal Trajectory Planning for Cylindrical Surfaces

The trajectory of the dip coating on cylindrical surfaces can be divided into three situations: reverse-dispersion, reverse-concentration and direction-parallel, which is shown in figure 3.

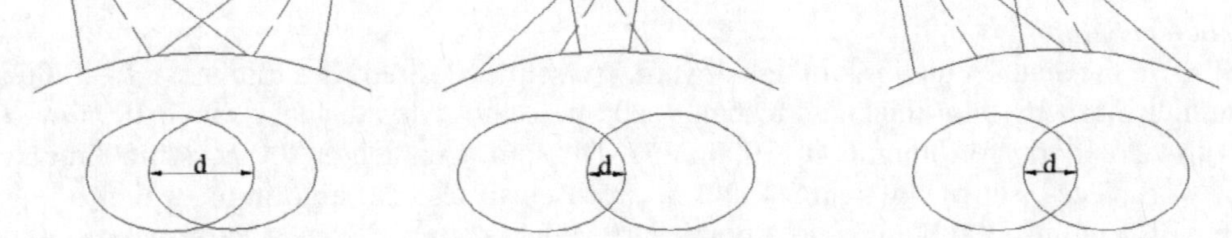

Figure 3. Three schemes of dip angle spray on cylindrical surfaces.

We employ figure 3(a) and 3(b) to calculate the painting thickness on a point s, figure 3(c) can also be calculated similarly. The painting thickness of a given point s can be expressed as follow:

$$T_s(x) = \begin{cases} T_1(x) & 0 < x < d - a(\pm\beta) \\ T_1(x) + T_2(x) & d - a(\pm\beta) < x < a(\pm\beta) \\ T_2(x) & a(\pm\beta) < x < d \end{cases} \tag{6}$$

Where a is the length of the long axes of dip angle spray model of the cylindrical surface; x is the position coordinate; T_1 and T_2 are dynamic single stroke coating deposition thickness which is calculated from static coating growth rate model:

$$T_1(x)=2\int_0^{t_1} f(r_1,x,y,\pm\beta)dt,\ \ 0<x<a(\pm\beta);\ \ T_2(x)=2\int_0^{t_2} f(r_2,x,y,\mp\beta)dt,\ \ d-a(\pm\beta)<x<d. \quad (7)$$

Where

$$t_1=b\sqrt{1-(\frac{x\mp\Delta a}{a})^2}/v;t_2=b\sqrt{1-(\frac{x-d\pm\Delta a}{a})^2}/v$$

$$r_1=\frac{hsin\alpha\sqrt{x^2+y^2}}{h\pm x\sin\beta};r_2=\frac{hsin\alpha\sqrt{(x-d\pm\Delta a)^2+y^2}}{h\mp(x-d)\sin\beta}$$

$$a(\pm\beta)=\frac{K_1(\pm\beta)\sin\varphi}{\cos K_2(\pm\beta)};\varphi=\arctan(\frac{R}{h});\ y=vt_{1,2};\Delta a=|a-a(\pm\beta)| \quad (8)$$

$$K_1(\pm\beta)=(K_3-\sqrt{\rho^2-h^2\tan^2(\varphi\pm\varphi_1)-2\rho h\cos\beta\tan^2(\varphi\pm\varphi_1)})\cos(\varphi\pm\varphi_1)$$

$$K_2(\pm\beta)=\arcsin(\frac{K_3\sin(\varphi\pm\varphi_1)}{\rho})-\arcsin(\frac{h\sin(\varphi)}{\rho}+\sin(\varphi\pm\beta))$$

$$K_3=\sqrt{\rho^2+h^2+2\rho h\cos\beta};\varphi_1=\arcsin(\frac{\rho\sin\beta}{K_3});a=\frac{a(+\beta)+a(-\beta)}{2};b=\frac{aR}{\sqrt{a^2-\Delta a^2}}.$$

Note that: t_1 and t_2 are the spraying time on the point *s* by the two adjacent strokes respectively; r_1 and r_2 are the distance from point s to the projection center of the spray gun of two adjacent strokes; Δa is the vertical distance of the elliptical center to the spray trajectory; *b* is the length of the short axes of dip angle spray model of the cylindrical patch.

In order to find the optimal velocity *v*, the overlap distance *d* and the dip angle of the spray gun β, the mean square error of the thickness deviation from the average thickness T_d must be minimized. Here the coating uniformity is taken as the objective for trajectory optimization:

$$\min E=(T_d\text{-}T_{\max})^2+(T_d\text{-}T_{\min})^2. \quad (9)$$

The optimal parameters *d*, *v*, α can be obtained by the genetic algorithm.

4. Simulation Results

4.1. Problem Setup.

According to the actual situation of the shipyard, we set the desired film thickness T_d=100μm, and the maximum error of the film thickness to be Ts=20μm. The overlap distance range 100mm<*d*<500mm, spray gun velocity range 200mm/s<*v*<1000mm/s. The spray gun height 200mm<*h*<600mm.

The workpieces, shown in figure 4 and 5, are used to test the algorithm, which is a cylindrical surface with a radius of 1000mm and a horns surface in a shipyard consisted of two different patches. The first patch is a plane while the second patch is a cylindrical surface with a radius of 1500mm, the angle of the two patches is 120°. We assume that the dip angle of the spray gun for the plane to be β_1, then the dip angle for the cylindrical surface is 60°–β_1.

Figure 4. A cylindrical surface.

Figure 5. A horns surface.

4.2. Analysis of Results.

4.2.1. Dip angle spray on the cylindrical surface. We set the dip angle of the spray gun to be 10°, 20° and then use the function(9) to optimize parameters in three different conditions shown in figure 3, the distributions of painting thickness are shown in figure 6.

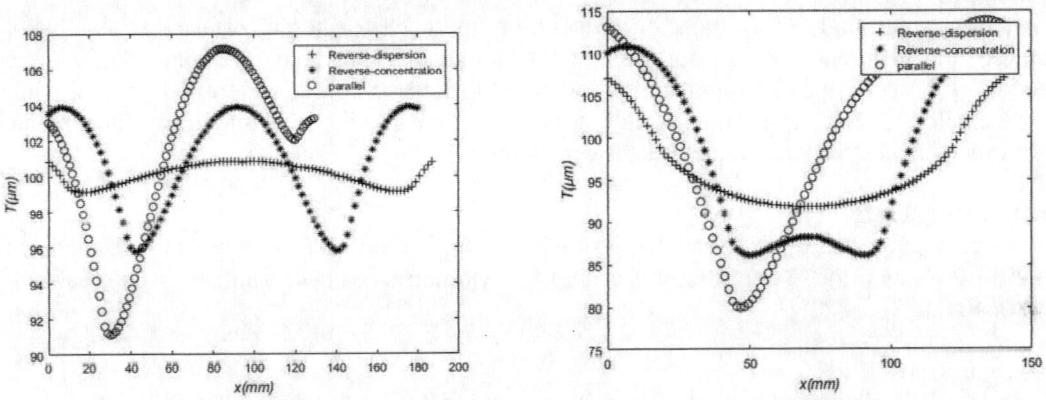

(a)10° (b)20°

Figure 6. The painting thickness distribution of different dip angles.

From the simulation results shown in figure 6(a), (b) we can conclude that the painting quality of reverse-dispersion on the cylindrical surface is the best.

4.2.2. Dip angle spray on the horns surface. The painting strategy on the horns surface is shown in the figure 7, we use reverse-dispersion to ensure the best painting quality on the horns surface.

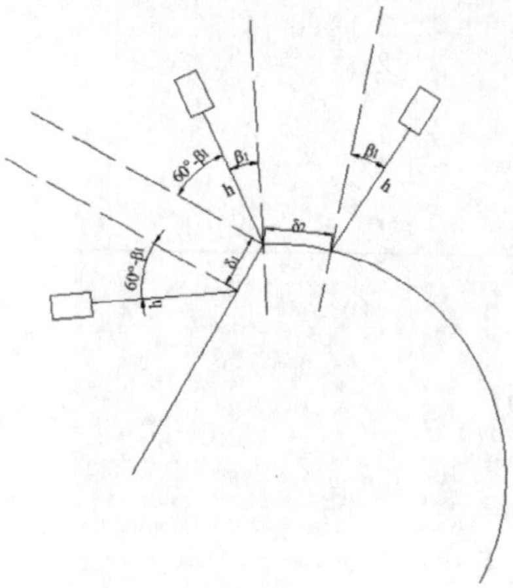

Figure 7. The painting strategy.

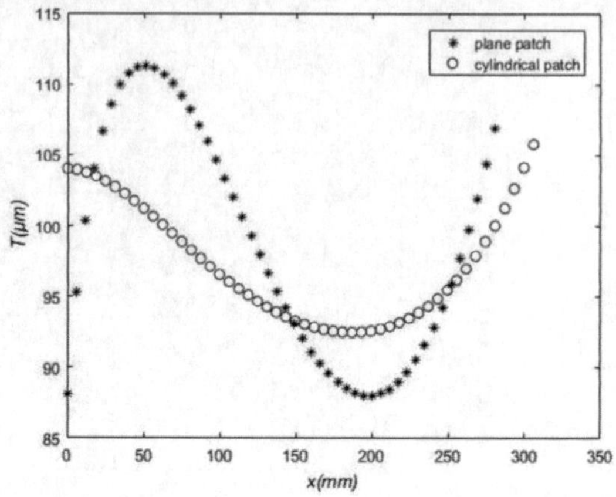

Figure 8. The painting thickness distribution.

The painting quality can be seen in figure 8, we can conclude that the painting strategy in this paper yields satisfactory results.

5. Conclusion

In this paper, a dip angle spray trajectory generation approach for horns surfaces is developed. By examining three different dip angle spray strategies on cylindrical surfaces, we conclude that reverse-dispersion yields better painting uniformity compared to reverse-concentration and parallel painting. Then we employ reverse-dispersion strategy on horns surface and it can be seen from the simulations that our approach is effective and yields satisfactory results.

6. Acknowledgments

This project is supported by the Shanghai Municipal Commission of Economy and Information (No. CXY-2013-25), and the Science and Technology Commission of Shanghai Municipality (No. 14111104400).

7. References

[1] Bo Z, Xi Z, Meng Z, et al. Off-line Programming System of Industrial Robot for Spraying Manufacturing Optimization[C]// Control Conference. IEEE, 2014:8495-8500.

[2] Yan C, Hua Y, Wang L, et al. Coating uniformity with a uniform robotic spray gun velocity[J]. Journal of Tsinghua University, 2010, 50(8):1210-1213+1218.

[3] Liu Y, Teng J, Song W. Trajectory Planning and Optimal Method of Spray Manipulator for Furniture Surface Painting[C]// IEEE Conference on Industrial Electronics and Applications. IEEE, 2007:2030-2035.

[4] Barros J R M D. Application research of the spraying trajectory optimization in-plane for the new model of spray gun[J]. Machinery Design & Manufacture, 2010.

[5] Shu-Mei MA, Xi Luo, Ai-Ping LI, et al. Painting parametric optimization using AMPSO algorithm on ship painting robots[J]. Chinese Journal of Construction Machinery, 2016.

[6] Tuna Balkan, M.A. Sahir Arikan. Modeling of paint flow rate flux for circular paint sprays by using experimental paint thickness distribution[J]. Mechanics Research Communications, 2006, 26(1):609-617.

[7] Zaki A, Eskander M. Spray painting of a general three-dimensional surface[C]// Ieee/rsj International Conference on Intelligent Robots and Systems. IEEE, 2000:2172-2177 vol.3.

[8] Chen H, Sheng W, Xi N, et al. Automated robot trajectory planning for spray painting of free-form surfaces in automotive manufacturing[C]// IEEE International Conference on Robotics and Automation, 2002. Proceedings. ICRA. IEEE, 2002:450-455 vol.1.

[9] Sheng W, Chen H, Xi N, et al. Tool path planning for compound surfaces in spray forming processes[J]. IEEE Transactions on Automation Science & Engineering, 2005, 2(3):240-249.
[10] Chen H, Xi N, Sheng W, et al. Optimizing material distribution for tool trajectory generation in surface manufacturing[C]// Ieee/asme International Conference on Advanced Intelligent Mechatronics. Proceedings. IEEE Xplore, 2005:1389-1394.
[11] Heping Chen, Ning Xi, Weihua Sheng, et al. General Framework of Optimal Tool Trajectory Planning for Free-Form Surfaces in Surface Manufacturing[J]. Journal of Manufacturing Science & Engineering, 2005, 127(1):49-59.
[12] Gong D. Trajectory Optimization of Spray Painting Robot for Natural Quadric Surfaces[J]. China Mechanical Engineering, 2011, 22(3):282-290.
[13] Zhang P, Gong J, Wei L. Arc Spray Gun New Models of Considering the Flow in the Process of Spraying for Spray-Painting Robot[C]// International Conference on Intelligent Human-Machine Systems and Cybernetics. IEEE, 2015:165-168.
[14] Chen W, Tang Y, Zhao Q. A novel trajectory planning scheme for spray painting robot with Bézier curves[C]// Control and Decision Conference. IEEE, 2016:6746-6750.
[15] Chen W, Sun C, Liu H, et al. Path planning scheme for spray painting robot with Bézier curves on complex curved surfaces[C]// Automation. IEEE, 2017:698-703.
[16] Chen Wei. Research on Key Techniques of Robotic Spray Painting Trajectory Optimization[D]. Zhenjiang: Jiangsu University, 2013.
[17] Fu Z, Xiao B, Wu C, et al. A genetic algorithm-based surface segmentation method for spray painting robotics[C]// Control and Decision Conference. IEEE, 2017:4049-4054.
[18] Zhou B, Fang F, Shao Z, et al. Fast and templatable path planning of spray painting robots for regular surfaces[C]// Control Conference. IEEE, 2015:5925-5930.
[19] Zeng Y, Gong J, XU Ning, et al. Tool trajectory optimization of spray painting robot for horns surface[J]. Journal of Zhejiang University, 2014, 48(5):791-798.
[20] Zeng Y, Gong J, XU Ning, et al. The Dip Angle Spray Trajectory Optimization Method Research[J]. Machinery Design & Manufacture, 2014.
[21] Zeng Y, Zhang Y, He J, et al. Prediction Model of Coating Growth Rate for Varied Dip-Angle Spraying Based on Gaussian Sum Model[J]. Mathematical Problems in Engineering, 2016, (2016-12-6), 2016, 2016(4):1-7.

IOP Conf. Series: Materials Science and Engineering **521** (2019) 012016 doi:10.1088/1757-899X/521/1/012016

Improvement of the fretting wear resistance of Ti6Al4V by application of hydrostatic ball burnishing.

S Swirad[1]

[1]Rzeszow University of Technology, The Faculty of Mechanical Engineering and Aeronautics, al. Powstancow Warszawy 12, 35-959 Rzeszow, Poland

s.swirad@prz.edu.pl

Abstract. The objective of the research was to improve the fretting wear resistance of titanium alloy by hydrostatic ball burnishing. Ball burnishing is an economic surface finishing process for wide range of components. It is based on pressing and rolling a hydraulically floated ball along the surface of the component being treated, to plastically deform surface asperities. This process generates a deep field of compressive residual stresses (up to 1.0 mm) in the outer layers of the component. The compressive residual stresses and the high quality of the surface finish improve the wear resistance of the part. Experiments were made using ball-on-disc tribotester in oscillatory motion. 100Cr6 sphere co-acted with a disc made of titanium alloy Ti–6Al–4V. The friction force was monitored as a function of time during the test. Wear of disc was measured after the test using the non-contact 3D surface profiler. The optimal ball burnishing parameters have been determined after conducting the Taguchi L9 matrix experiment. During tests, normal load was kept constant at 50 N within the contact with frequency of 50 Hz and displacement amplitude of 0.02 mm. The results confirmed that hydrostatic ball burnishing had a significant influence on friction and wear.

1. Introduction

Fretting is defined as a small amplitude oscillatory movement which occurs between two surfaces in contact. Partial slip or gross slip condition can be observed in the fretting contact. For the partial slip case there are parts of contact where no slip displacement occurs, but in gross slip case there is slip displacement across the whole contact [1]. The fretting regimes can be described as functions of normal load and displacements. This oscillatory movement often results in the formation of oxide debris and thus, is called "fretting wear" or "fretting corrosion". In most cases, the cause of this movement is due to one of the members being placed under cyclic fatigue loading. The case where a fatigue loading is present is more damaging and is called "fretting fatigue". The type of damage, wear or fatigue is related also to the sliding condition, i.e. partial slip or gross slip. The fretting fatigue is related with higher loads, a smaller amplitude of motion and contact within partial slip regime. Wear is leading in the gross slip regime [2]. Fretting fatigue is a common problem which has detrimental affects on a broad range of engineering parts. Several factors can affect crack nucleation under fretting fatigue. These factors are: the loading condition, the surface contact condition, and the operating environment. The relevant loading condition includes the contact slip amplitude normal contact pressure, and the bulk applied stress amplitude and mean stress. The surface contact condition includes the surface roughness, chemical composition of the contact region, residual surface stress and any coatings or lubricants on the surface.

Surface treatments have been shown to greatly improve the overall life of engineering components that undergo fretting wear. There are three major principles behind the types of treatments that are commonly used in cases of dry friction where regular lubricants cannot be maintained. First of all is to apply a low friction coating on the contact surface. Second is to apply a soft coating that easily shears away, creating a protective layer between the surface s of the material. The finale is to generate a compressive stress in the surface layer which resists the initiation and growth of the cracks. With a lower coefficient of friction between two surfaces the frictional force between the components is reduced, thus decreasing the amount of damage. This can be achieved by applying hard coatings such as TiN or diamond coatings. This method is most effective for the condition where gross slip is present. The different method to improve fretting wear is to apply a soft coating on the components that can easily shear. The resulting debris that is created from shearing forms a layer between the contacting surfaces preventing abrasive wear and initial welding which can occur during fretting. The debris can help decrease the coefficient of friction which also improves the fretting fatigue strength. The final major family of surface treatments involves a compressive stress in the surface layer, typically through a cold working process, which helps improve the fretting wear. The common cold working processes that can be employed in surface treatment application are shot peening, laser shock peening, and low plasticity burnishing. The principle in these processes is that it creates a compressive layer in the surface layer of the material that delays fretting wear [3].

Burnishing is an acknowledged method of mechanical surface treatment, to impart specific mechanical, physical and tribological properties. In low plasticity burnishing a smooth, free rolling spherical ball under a normal force makes a single pass over the surface of material leaving a residual compressive stress in the surface. Burnishing process is capable of improving resistance to wear, oxidation and corrosion. These enhancements can be extended to minimize friction and reduce adhesion [4,5].

Due to its excellent combination of toughness, strength and corrosion resistance, Ti6Al4V is one of the most widely used titanium alloys. However, Ti6Al4V alloy is known to possess poor wear resistance that restricts its application particularly in areas involving wear and friction. [6,7] . Many scientists studied the dry sliding wear behavior of Ti6Al4V alloy in sliding condition at different sliding velocity and applied loads to confirm the poor wear resistance of the alloy plastic deformation and low protection exerted by the surface oxidation. One of the approaches for enhancing the tribological properties of titanium alloys for improved fretting resistance is to apply surface modification for increased hardness and reduction in coefficient of friction [8,9,10]. In this study, the effect of machining parameters on the wear volume and coefficient of friction in the ball burnishing process of Ti6Al4V alloy were investigated.

2. Experimental details

Experiments were made using a ball-on-disc tribotester Optimol SRV in oscillatory motion. A steel ball from 100Cr6 of 60 HRC hardness and 10 mm diameter contacted a disc made of Ti6Al4V alloy under dry fretting condition. During the test, coefficient of friction ratio being the of the tangential force and the normal load was monitored as a function of time. Wear of the disc was measured by the white light interferometer Talysurf CCI Lite. During tests, normal load was kept constant at 50 N within the contact with frequency of 50 Hz and displacement stroke of 0.1 mm. Duration of each test was 15 min. In the present study, three parameters, namely, burnishing force, velocity and width were identified and the range for each of the process parameters was determined through the preliminary experiments [11]. The burnishing system used in this research is equipped with a high-pressure hydraulic pump and the 6 mm diameter burnishing tool, both connected via high-pressure hoses. Pressure was supplied by the hydraulic pump, which is equipped with its own coolant tank and pump at the coolant maximum pressure of 40 MPa. The process parameters and their levels in the current investigation of burnishing process are summarized in table 1. According to Taguchi design concept, L_9, orthogonal array was selected [12]. Each burnishing parameters is assigned to a column and nine different process parameters. The experimental matrix for the current investigation using L_9 array is shown in table 1.

Table 1. Formatting sections, subsections and subsubsections.

Expt. No.	Coded values			Actual values			Wear volume	Signal to noise ratio S/N	Avg. coefficient of friction - COF	Signal to noise ratio S/N
	A	B	C	P [MPa]	v [mm/min]	a [mm]	V [mm³]	S/N V	COF	S/N COF
1	1	1	1	10	1000	0.04	0.017234	35.272	0.965	0.309
2	1	2	2	10	1500	0.06	0.017543	35.118	0.911	0.810
3	1	3	3	10	2000	0.08	0.017760	35.011	0.954	0.409
4	2	1	2	20	1000	0.06	0.016978	35.402	0.968	0.282
5	2	2	3	20	1500	0.08	0.016720	35.535	0.956	0.391
6	2	3	1	20	2000	0.04	0.016550	35.624	0.976	0.211
7	3	1	3	30	1000	0.08	0.015340	36.283	0.997	0.026
8	3	2	1	30	1500	0.04	0.014930	36.519	0.923	0.696
9	3	3	2	30	2000	0.06	0.014640	36.689	0.914	0.781
Surface before burnishing							0.017804		1.014	

3. Results and discussion

In the present study, the objective is to minimize the specific wear rate and confection of friction. Therefore, "smaller is better" category for the specific wear volume (V) and coefficient of friction (COF) has been selected. The S/N ratio associated with the objective function for each trial of the orthogonal array is given by:

$$\eta = -10 \times \log_{10}[(1/n) \times \sum \left(y_i^2\right)] \tag{1}$$

where n is the number of measurements in a trial/row and yi is the measured value in a run/row.
The corresponding S/N ratio for each experiment of L_9 orthogonal array were determined using eq. 1 and are presented in table 1. The means analysis based on S/N ratio was carried out to determine the optimal levels of process parameters. The results of analysis (for S/N ratio and means) for specific wear volume and coefficient of friction are presented in table 2 and 3, respectively. The level of the parameter with the highest value of S/N ratio is the best combination level. The optimal parameter setting is found to be A3, C1 and B3 for minimum specific wear volume. For the optimal parameter setting for minimum coefficient of friction is found to be A3, C1and B3. From the analysis of table 2 it is evident that the strongest influence on specific wear volume was exerted by the burnishing force (rank 1), burnishing width (rank 2) and burnishing velocity (rank 3). From figure 1, it is evident that wear volume is reduced or an increased in wear resistance is observed with the increase in burnishing force, and burnishing speed, whereas the wear volume decrease with increase burnishing width. This is due to the fact that as the burnishing force increased from 10 MPa to 30 MPa, hardness of the surface increased [13]. Further burnishing force increase compressive residual stress at the surface layer, thus hindering the growth of wear delamination.
From the analysis as shown in table 3, it is evident that the strongest influence on the coefficient of friction was exerted by the burnishing speed and burnishing width. Figure 2 illustrates the S/N ratio and main effect plot of burnishing parameters on coefficient of friction.
One can see from the analysis of figure 3a and 3b that build-ups with compacted debris were formed inside the disc scars, which is connected with higher wear of balls in corresponding places. The diameter of the wear scar in the burnished material is shorter than that in the unmodified material. Also, the depth of the wear scar in the modified material is shallower than that in the unmodified material.

Table 2. Response table for wear volume - S/N ratios (left side) and for the means (right side).

Level	Force [MPa]	Velocity [mm/min]	Width [mm]	Level	Force [MPa]	Velocity [mm/min]	Width [mm]
1	35.13	35.65	35.81	1	0.01751	0.01652	0.01624
2	35.52	35.72	35.74	2	0.01675	0.01640	0.01639
3	36.50	35.77	35.61	3	0.01497	0.01632	0.01661
Delta	1.36	0.12	0.20	Delta	0.00254	0.00020	0.00037
Rank	1	3	2	Rank	1	3	2

Table 3. Response table for coefficient of friction - S/N ratios (left side) and for the means (right side).

Level	Force [MPa]	Velocity [mm/min]	Width [mm]	Level	Force [MPa]	Velocity [mm/min]	Width [mm]
1	0.5094	0.2060	0.4055	1	0.9433	0.9767	0.9547
2	0.2948	0.6321	0.6244	2	0.9667	0.9300	0.9310
3	0.5010	0.4670	0.2753	3	0.9447	0.9480	0.9690
Delta	0.2146	0.4261	0.3491	Delta	0.0233	0.0467	0.0380
Rank	3	1	2	Rank	3	1	2

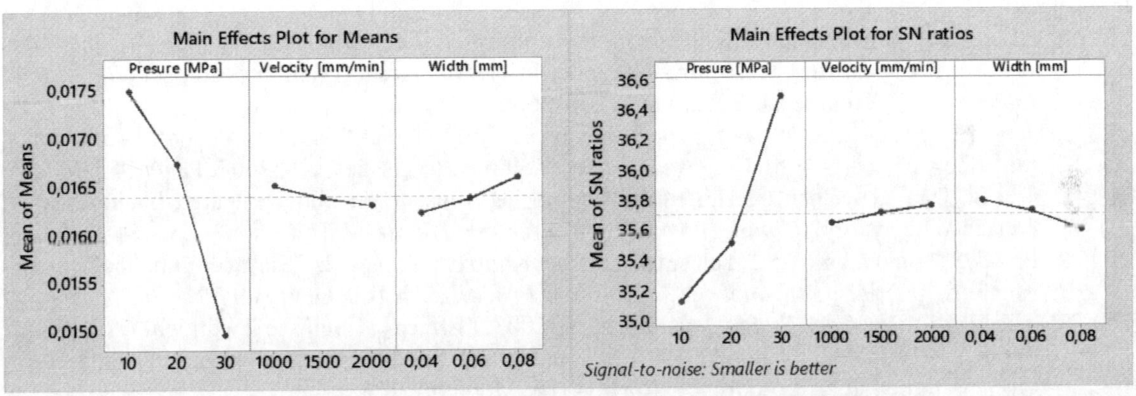

Figure 1. Main effect plot and S/N ratio for wear volume.

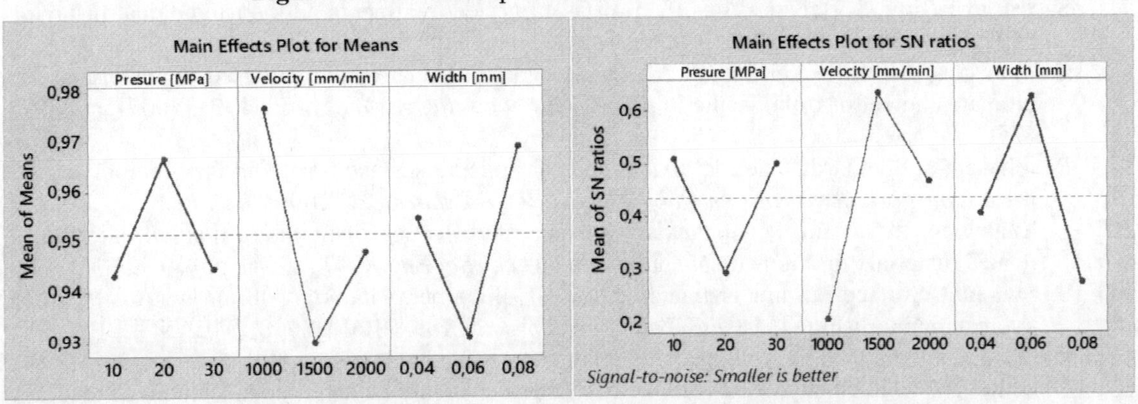

Figure 2. Main effect plot and S/N ratio for coefficient of friction .

IOP Conf. Series: Materials Science and Engineering **521** (2019) 012016 doi:10.1088/1757-899X/521/1/012016

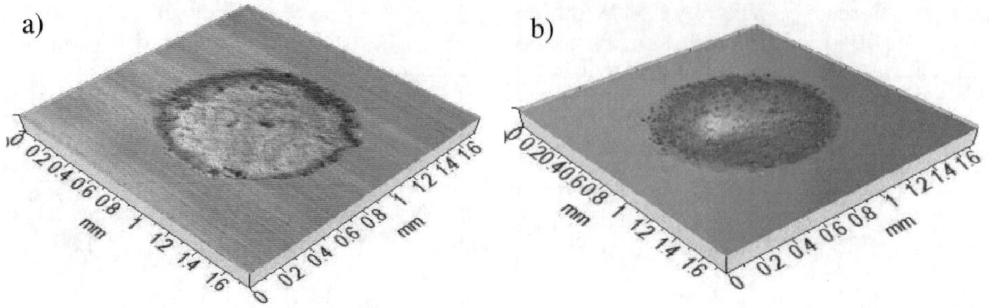

Figure 3. Isometric view of wear scars (a) before ball burnishing and (b) after ball burnishing.

4. Conclusion

Several fretting wear tests consisting Ti6Al4V on Ti6Al4V contacts were conducted. A combination of burnishing force in the high range and low burnishing width is helpful to minimize wear volume. However, medium range of burnishing force, speed and width could minimize the coefficient of friction. It is evident that specific, wear volume is reduced with the increase burnishing force. The burnishing force an burnishing width have major effect in minimizing the specific wear volume. Contrary the burnishing velocity have key role in minimizing the coefficient of friction.
Low plasticity burnishing treatment appears to be effective way to induce a compressive residual stress that inhibits fretting wear. The surface quality and tribological characteristics if titanium alloy Ti6Al4V can be improved by ball burnishing showing improved friction and wear performance.

References

[1] J.D. Lemm, A.R. Warmuth, S.R. Pearson, P.H. Shipway, The influence of surface hardness on the fretting wear of steel pairs—Its role in debris retention in the contact, *Tribology International*, Volume **81**, 2015, Pages 258-266.

[2] J. Ding, I.R. McColl, S.B. Leen, P.H. Shipway, A finite element based approach to simulating the effects of debris on fretting wear, Wear, Volume 263, Issues 1–6, 2007, Pages 481-491,

[3] D. Liu,B. Tang, H. Chen, J. He, MoS2 composite films on Ti alloy prepared by ion-beam-enhanced deposition, *Chinese Journal of Nonferrous Metals*, Vol.**11**, No. 3, pp. 454-460

[4] P. Prevey, P.S., Gabb, T.P., Telesman, J. and Kantzos, P., 2002, "Surface Enhancement of Metallic Materials," *Advanced Materials and Processeses*, **160(1)**, pp. 69-72.

[5] Shepard, M.J., Prevey, P.S. and Jayaraman, N, 2003, "Effects of Surface Treatment on Fretting Fatigue Perfomance of Ti-6Al-4V," Proceedings of the 8th National Turbine Engine High Cycle Fatigue (HCF) Conference, April 14-16, Monterey, CA.

[6] K.G. Budinski, Tribological properties of titanium alloys , *Wear*, **151** 1991, pp. 203-217

[7] S. Yerramareddy, S. BahadurThe effect of laser surface treatments on the tribological behavior of Ti–6Al–4V, *Wear*, **157** (1992), pp. 245-262

[8] G. Manisavagam, K.S. Amritpreet, A. Rajamanickam, K.G. Ashok, Ti based biomaterials: the ultimate choice for orthopaedic implants – a review *Prog Mater Sci*, **54 (3)** (2009), pp. 397-425

[9] R. Johnson, J. Eberhardt, Thermal oxidation: a promising surface treatment for titanium engine parts *Transportation Program Oak Ridge National Laboratory* (2006)

[10] L. Xuanyong, K.C. Paul, D. Chuanxian, Surface modification of titanium, titanium alloys and related materials for biomedical application, *Mater Sci Eng R*, **47** (2004), pp. 49-121

[11] S. Swirad - Surface texture characterization of alpha-beta titanium alloy (Ti6-Al4-V) after hydrostatic burnishing, *MATEC Web of Conferences* **189**, 01001 (2018) ,MEAMT 2018

[12] P.J. Ross, Taguchi techniques for quality engineering (2 ed.), McGraw-Hill, New York (1996)

[13] G. D. Revankar, R. Shetty, S. S. Rao, V. N. Gaitonde, Wear resistance enhancement of titanium alloy (Ti–6Al–4V) by ball burnishing process, *Journal of Materials Research and Technology*, **6**, 2017, Pages 13-32,

IOP Conf. Series: Materials Science and Engineering **521** (2019) 012017 doi:10.1088/1757-899X/521/1/012017

Survey: The Evolution of the Usage of Augmented Reality in Industry 4.0

Abir Gallala[1,a], Bassem Hichri[1] and Peter Plapper[1]

[1]Research Unit in Engineering Sciences University of Luxembourg, Luxembourg

[a]Corresponding author: abir.gallala@uni.lu

Abstract. The usage of Augmented Reality (AR) in industrial and modern manufacturing is more and more growing since the fourth industrial revolution. Using AR boost the digitization of the industrial production lines, gain time and money and improve maintenance tasks as well as the human-machine interaction. This paper is a literature review of the use of AR in industries including the use cases in different type of application such as design, simulation, maintenance, remote assistance, human-robot interaction and robot programming.

1. Introduction

Nowadays, the industrial landscape is being transformed its fourth revolution thanks to the rise of the internet of things, collaborative/autonomous robots, cyber-physical systems, internet of services. This transformation, also known as Industry 4.0, is the new area where the internet of things interconnects with cyber-physical systems. It makes a mixture of software, sensor, processor and communication technologies. Thus, it plays a huge role in making objects able to feed information into the network and eventually adds value to manufacturing processes. Industry 4.0 ultimately aims to construct an open and smart manufacturing platform for industrial-networked information applications based on a range of technologies cited in Figure 1. This concept is best described as "smart factory" which aims to merge virtual and real worlds together [2].

Industrial robots play an important role in the Industry 4.0. They have been involved considerably in the last decades. They are becoming more collaborative, productive, flexible, safer, and versatile and they have a huge capability to interact with each other and work safely side by side with humans. Programming industrial robots in large manufacturing, as well as small and medium-sized enterprises, requires a lot of time. Thus, a new method of programming industrial robots that came out with this revolution is Augmented Reality (AR).

AR is a way of transforming and visualizing digital information as pictures or videos. AR has been used for industrial tasks for more than 20 years [3], e.g. it can be used only for visualization of information or even for interaction. With the last few years, AR was involved widely in different fields and activities.

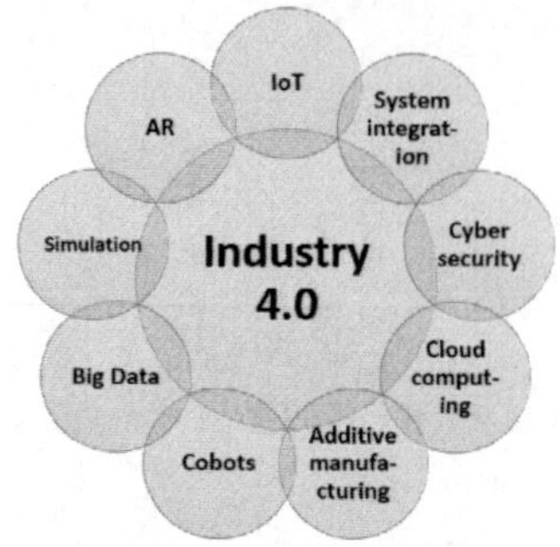

Figure 1. Technologies related to industry 4.0 [1].

Notably, it is being involved in education by enhancing the student's motivation to learn [4] and training teachers to become more effective [5]. It also boosts the digitization of the industry production lines or maintenance tasks [6, 7].

This paper reviewed AR-based industrial applications that appears alongside the fourth industrial revolution such as maintenance, assembly, remote assistance as well as human-robot interaction and robot programming. It is organized as follows. Section 2 is about the general usage of AR in industry 4.0 including the definition of AR and the selection criteria. Section 3 summarizes the usage of AR for visualization applications such as simulation, design, maintenance, assembly and disassembly. Section 4 discussed remote communication usage such as remote assistance and remote maintenance. Section 5 analyzed AR usage for robotics: human-robot interaction (HRI), robot manipulation and robot path planning. Finally, the conclusion in section 6.

2. Usage of Augmented Reality in industry 4.0

All along the appearance of industry 4.0, digitalization and smart factories, many manufacturers start exploring the benefits that AR can offer in a smart industrial environment.

AR is the integration of virtual objects in the real world [8]. Therefore, it enhances the reality by appearing virtual objects alongside the real ones. Professor Paul Milgram is the first who introduced the concept of the reality-virtuality. Mixed Reality Continuum, shown in Figure 2., is a one-dimensional array between the Real and the Virtual Environments [9].

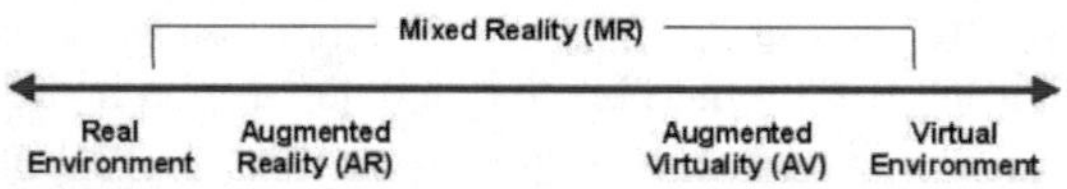

Figure 2. Milgrams Mixed Reality Continuum [9].

After that, as a compliment of Milgram's Mixed Reality, Mann [10] presented the concept of Mediated Reality shown in Fig. 3 for modifying the real view instead of just adding to it.

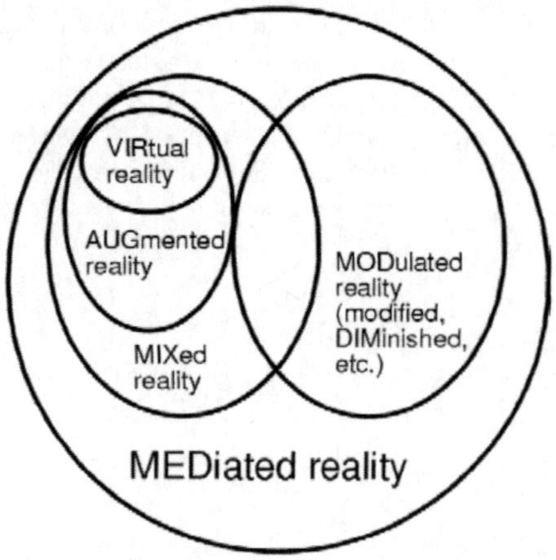

Figure 3. Manns Mediated Reality [10].

2.1 Selection criteria
This review includes the recent projects that used AR in industry 4.0. In the last period, the usage of AR in the smart factories has involved and took a different kind of interventions. Before 2014, most of the use cases of AR was for data visualization either to assist the employees [11] or to improve maintenance tasks [12]. After that, it was developed to assist employees thanks to the remote assistance and/or maintenance [13]. Furthermore, it starts to be used for the human-robot interaction [14] and collaboration, robot programming [15] and robot trajectories planning [8]. Figure 4. illustrates how AR is being implemented in industry 4.0 based on analyzing most of the projects realized since 2012.

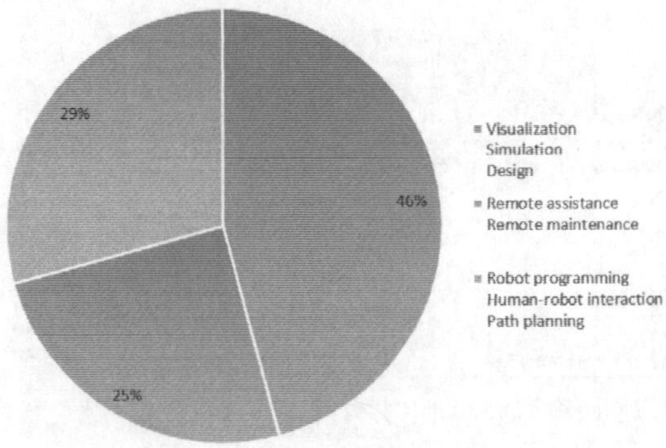

Figure 4. AR implementation in industry 4.0.

3. Visualization usage

3.1. Data visualization, simulation and design
Data displaying and simulation is one of the basic application of AR in general and in manufacturing in special. As its name said, it is just projection of information from a database to a user interface for different usage.
In 2012 A. Nee sited in [16] a list of design applications using AR in addition to some other industrial applications and then a year after, J. Novak-Marcincin [11] suggested a similar model that enables a user to visualize information related to the manufacturing process, additional text and acoustic information related to the work environment, conditions and objects.
In 2014, HC Fang presented in [14] an HRI AR-based interface that aims to assist users when interacting with a virtual robot in a real work environment. He adopted for that an AR Tool Kit-based hand-held device tracking method which was a marker-cube attached with a probe and virtual robot registration. It allows users to interact with spatial information of the working environment. It can be used to guide the virtual robot to intervene in the path planning and end-effector (EE) orientation planning processes. The actual working environment, the virtual robot model, the trajectory information, as well as the interaction processes are visualized through a monitor-based display. In the same year, V. Paelke also presented a similar application [17].
Another concrete system was presented in [18] in the same year. Lofvendah worked together with ABB to build an android application that displays an ABB industrial robots SafeMove coordinates using AR [19]. It was designed to help engineers verify the SafeMove settings. First, a user enters the coordinates in SafeMove that will create an area and then create its virtual model. Then, they need to track the robot in order to place the virtual model of the safety zone. They placed several fiducial markers[20] around the real robot to know where to place the model. A function was added to allow the adjustment of the safety zone. Next, a calibration is required to integrate the virtual model with the real one. Finally, the last communication between the robot and the application is needed. To use the application, a user have to hold a tablet and point it towards a robot then he will see on the screen the augmented objects with the virtual safety zone as well as the possible movements and reachable areas of the robot. In the end, they tested their application with ABB employees. They faced a problem by holding the device all the time. Thus a solution was based on replacing the tablet with a wearable device such an AR head-mounted device.
In [21], an AR interactive space has been presented in 2015 in Italy. It aims to visualize and interact with virtual objects and models.

IOP Conf. Series: Materials Science and Engineering **521** (2019) 012017 doi:10.1088/1757-899X/521/1/012017

In the same context, a CNC aid machine was developed by J. Chardonnet based on AR [22] in 2017. It projects on a tablet or wearable device personalized information about process status and a real-time 3D simulation of the machine. S. Suto [23] has presented in the same year an AR-based simulation application for human-robot collaboration. His main idea is to digitize the human gestures using a Kinect sensor to simulate the collision in the virtual space in order to avoid it in the real one. Furthermore, it is also able to generate a collision-free trajectory for the robot.

As a common application, D. Mourtzis [24] has developed recently an AR-based computer aided manufacturing system that delivers informative texts, safety zones in the area, step-by-step instructions and some other indications as shown in Figure 5.

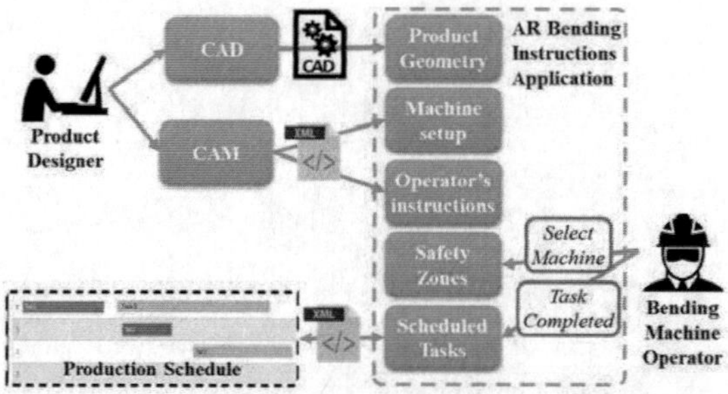

Figure 5. The architecture of Mourtzis's proposed approach [24].

This system promotes the efficiency of considering and resolving appearing problems.

A design and touch-based interaction interface was pre- sented also in 2018 in [25] to enhance the human-machine interaction. The prototype enables users to design their own object's shapes and products to be transferred to a G-code and then printed.

This application, as well as all presented prototypes and systems, prove the performance of this technology on making the simulation of models, visualization of virtual objects and the interaction with them as easy as it is real.

3.2 Maintenance

Maintenance counts around 60 to 70 % from the core activity of the production life-cycle [12]. Since unexpected machine breakdown could be time-consuming and this can directly influence the productivity, AR has arisen as an enabling technology for handling the increasingly complex maintenance processes.

R. Palmarini [26] has distinguished in his systematic literature review (SLR) several implementations of AR in maintenance applications. Figure 6. represents the fields of the application being used AR in maintenance throughout 30 studies. Dini [27] also came out with the same results. The mechanical field is the highest area of application, as mentioned in Figure below, and it includes the maintenance activities in different sectors such as aviation, automotive and military... In the same SLR, Palmarini has mentioned as well the percentage of the maintenance operations which are: 33% of dis-assembly and assembly operations, 26% repair activities, 26% Inspection and diagnosis and 15% training operations.

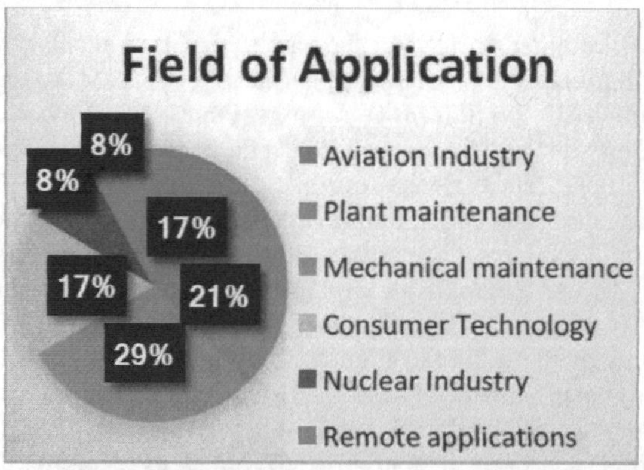

Figure 6. Field of application of AR for maintenance reported by R. Palmarini [26].

Later, he presented an AR-based method for maintenance [28]. His innovative system aims to support the maintenance tasks by guiding the employee to determine the requirements and the constraints for every different operation. The method is to develop a list of questions that helps for the point directly on the problem.
Webel defines training as the process that aims to transfer maintenance skills to technicians [29]. He developed a novel method for multimodal AR-based training of maintenance skills.
Mourtzis proposed in [30] a Product-Service System platform for tele-maintenance support using AR. This platform enables technicians and manufacturer experts to communicate maintenance instructions and exchange data and feedback reports. Figure 7. details the proposed architecture of this solution. It is composed by three steps that are executed every time which are:(i) malfunction report composition, (ii) diagnosis and AR maintenance instruction generation and, (iii) mainte- nance and evaluation. First, the system registers a report of the set of features for a scheduled maintenance or unexpected malfunction and then send it to the maintenance support provider. After viewing the report, it creates an enhanced failure report containing a cloud-based feedback mechanism to be sent to the manufacturer in a short time. If needed, the expert assists the technicians by giving them AR instructions. The hardware design of their system uses three main devices: a set of optical see-through Augmented Reality goggles, a laptop PC and a mobile device [31].

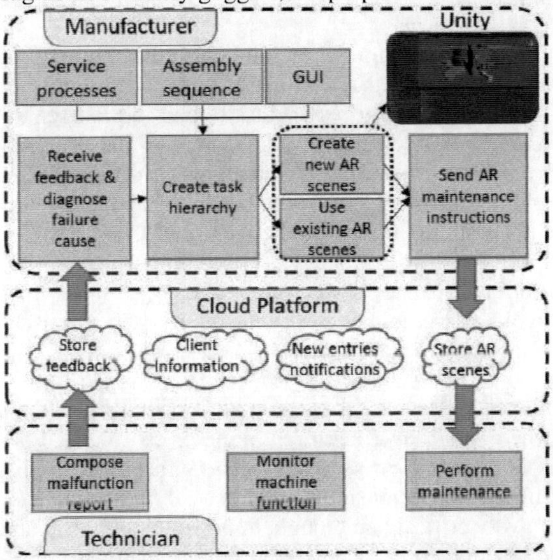

Figure 7. Architecture of the developed solution by Mourtzis.

For a hand-free vision system, they used as an AR head-mounted device. Then, the computer is the element responsible for executing the AR application and the communication handler. The third element is the mobile device that considers the interface of communication between the operator and the AR application.
Many examples and utilization of AR in maintenance could be pointed out. Previous literature [32-34] cited different applications of maintenance and repair tasks. All previous research and industrial projects prove the same result that the AR usage in maintenance reduce the machine breakdown time and improve the employee's abilities to fix and point faster on the problem.

3.3 Assembly/Dis-assembly

Assembly and/or disassembly are unavoidable and critical processes for any product's lifecycle. X. Wang [35] has published a very detailed review of AR assembly research. He defined assembly as the operation of grouping altogether some parts in order to form the desired product with good added value. Furthermore, he explained how AR resolved the two major problems in the assembly which are: 1) users need to identify technical specifications and constraints such as the mating surfaces, axes and edges which could be a very complex task, 2) 3D software does not help the user to interact directly with the parts neither manually manipulate them. However, AR came up to build up a pragmatic synergy between the user and the product by merging all together augmented and real objects along with making the manipulation of the pieces possible.
ARDIS is an AR-guided application for products dis-assembly proposed by M.Chang [36]. ARDIS generates the information automatically without any external/human intervention. Therefore, it makes the disassembly

process more intuitive and efficient. It proves that this application has a big impact on the re-manufacturing operation as well.

MotionEAP [37] is another AR industrial assembly interface developed in Germany. MotionEAP has many capabilities and options such as 1) It projects the instructions to the workers as well as all the steps, 2) It recognizes the worker's activities and the actual assembly step and 3) It is able to give visual feedback at different levels.

Another German AR assembly interface proposed by C.Kollatsch [31] and shown in Figure 8.

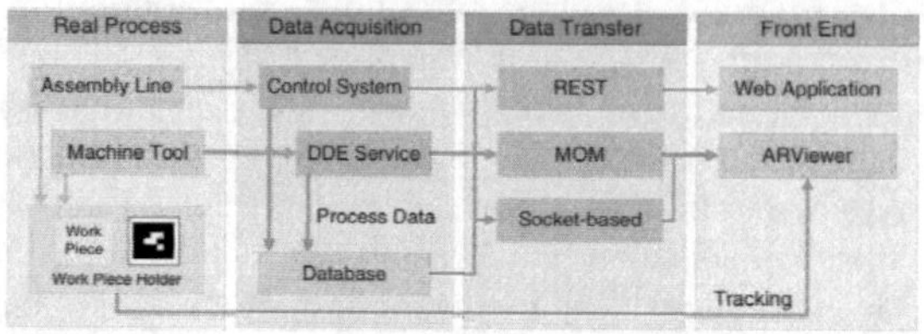

Figure 8. The monitoring concept of the AR-assembly applica- tion proposed Kollatsch [31].

It consists mainly of 4 stages. First, it starts from the real assembly line and real objects for the process data acquisition. After that, the system gathers all parts information and details to be first transferred and then stored in a database. Finally, is to project the necessary assistance to the user to accomplish the assembly operation. Kollatsch chose an Android device for the visualization of his assembly application. Most of the interviewed workers gave positive feedback for using such application during the assembly process.

Recently, Danielsson [38] has a similar approach but he replaced the handheld device by an AR head-mounted device that is an AR-glasses. In his work, he presented the results of the user's interview regarding the usage of this technology in an automotive assembly.

All presented AR-based systems offer assistance to the users all along of an assembly or disassembly process. It presents the proper information of an assembly process and step-by-step approach of the operation.

4. Remote assistance usage

The remote assistance, frequently called see-what-I-see remote cooperation, is the ability to hook up with an expert on-site anytime and anywhere. Whenever an urgent issue or complicated problem appears, workers in the workplace can easily connect with expertise from outside for a quick diagnose. This kind of AR application is a real-time shared interface in both sides.

Maintenance tasks in factories are very critical and though, thus they require fast and efficient interventions. D. Mourtzis [13] and M. Fiorentino [39] has roughly presented remote maintenance using AR. Then, Fiorentino made an evaluation of the efficiency of this aspect.

Almost alike, the ACAAR system presented in [40] enables a user to add digital information such as texts, pictures or computer-aided design (CAD) models. Furthermore, it can explain the contents regarding the required maintenance tasks.

F. Ferrise suggested in [41] an AR application for remote maintenance. It aims to enable the connection of an expert situated in a control room with an ordinary technician located on the promises.

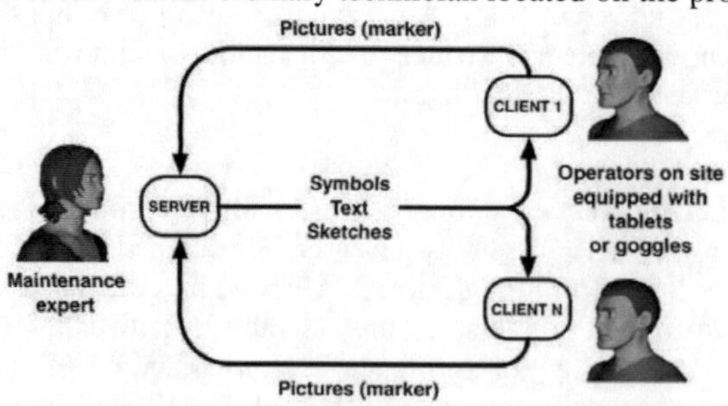

Figure 9. The system proposed by F. Ferrise [41].

The system presented in Figure 9. starts when one or more operator(s), called also a client, send pictures of the real scene to the expert, called server.

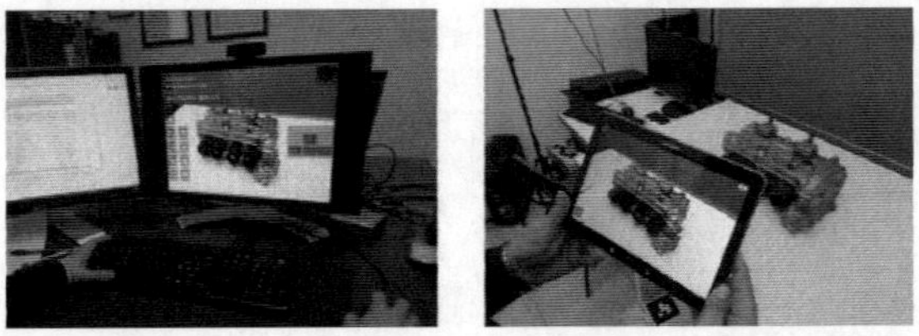

Figure 10. Real exemple of the system proposed by F. Ferrise [41].

As shown in the example in Figure 10., the expert indicates what and where the problem is by using symbols, text and some information.

In [42], a worker bears an AR device that records a real-time video stream of what he sees in addition to other data such as sensor information. The video has to be sent to an expert from outside and he can give an exact solution with additional information on real time, so he doesn't need to move to the factory to fix the problem.

It is been proved by all previous applications and industrial agents who tested this technology in their labs, that remote assistance reduced the machine downtime, saved money as well as time.

5. Robot programming usage

Programming industrial robots methods are increasing day after day in order to make it easier, faster and cheaper. Z. Pan detailed in his review in [15] the different programming methods for industrial robots. He discussed two of traditional programming methods: Online and Offline robot programming (OLP). During an online programming, the robot operator moves the robot manually to execute simple tasks. It's an efficient method but over-time it could not handle complex programs. OLP generates the programs using 3D CAD models. The operator can check and simulate the program before downloading it to the real robot such as safety zones, reachability and movements... it is even possible to use additional sensors in OLP. This method is more complex than the online programming and it requires more qualified engineers and more time. Recently, AR starts to be used for making robot programming that easy that any non-expert robot programmer would be able to setup any industrial application.

The main usage of AR on robot programming are HRI, robot manipulation and path/trajectory planning.

5.1 Human-Robot Interaction (HRI)

Industry 4.0 came up with the revolution of the human-robot collaboration and introduced collaborative robots where humans and robots work alongside together and also called "cobots". The cobots are becoming more adaptive and aware of their collaborators. This has come up with the research about the interaction and collaboration process. One of the most suggested and successful technologies was AR. It was used as: an interface for visualization and interaction like in [14], a third point of view for robot intentions visualizations as presented in [43], an application to support the HRI and cooperation such in [44] and [45], a teach pendant interface as defined in [46], or an interface for guiding users throughout an unfamiliar assembly operation just as demonstrated in [47].

Maly had an evaluation of the usage of AR interfaces for the human-robot interaction in [48]. After he developed an AR application for industrial robot visualization, highlighting parts of the robot and safe zone detection, he made an evaluation for his prototype. Six participants engaged in operating the system's features. All of them were satisfied with the data visualization and well confirmed that the hand gestures were fast except some gestures were not correctly detected which means the prototype wasn't 100% successful.

This section illustrated some AR-based HRI dedicated for visualization and human-robot collaboration operations but these are not its only application for robots. Recently, it started to be used for more than the only visualization but for robot programming as well.

5.2 Robot manipulation

The technological revolution aims to make objects more intelligent, thus human can easily communicate and deal with them. Using AR for robot manipulation is also aiming to make the human-robot interaction more easy even for non-expert operators and without the need of losing a lot of time in practicing how to move or to manipulate a robot especially with the important increase of numbers of cobots in factories. In 2016, Chryssolouris has proposed in [49] a robot programming method for robot manipulation using only body and hand gestures and without using any programming language.

The architecture of the system presented in Figure 11. lay-out the method used in this application.

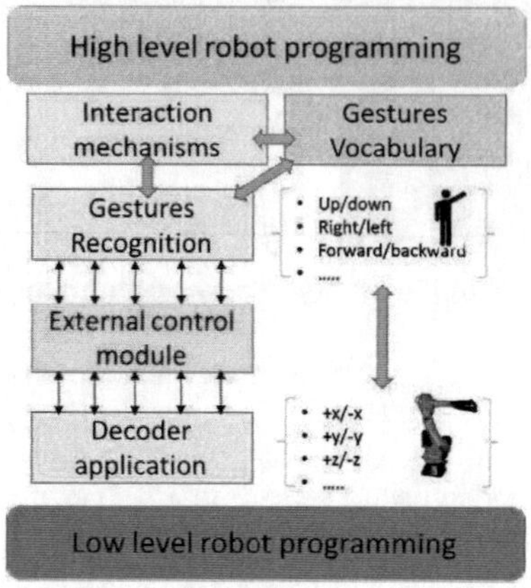

Figure 11. The system architecture proposed in [49].

The high-level robot programming is reached by using commands in the form of body and hand gestures vocabulary tracked by the intermediate of external device or sensors. The predefined gestures enable the user to control the robot movement on the 3 axis x, y and z. As an example, to move the robot on the direction +x, the user has to extend his right hand upwards and hold the left down. To move along the y-direction, the operator has to extend the right hand at the height of his shoulder and points to the left side. In the same year, another similar system was proposed in [50]. The idea is about an AR handheld device presenting a task-based interface for the robot control. They tested this approach with 19 industrial robot programmers where they confirm that it effectively reduced the mental demand of the programmers during the programming operation as well as required time.

A year after, L.Wang [51] worked on an application that virtually augments robot assembly commands and instructions. R.Matarneh [52] developed a voice control method for robotic assembly process models.

Recently in 2018, J.Wassermann [53] presented a com- pletely new approach which aims to manipulate a robot based on a cloud system. The main program generates the control instructions and then the AR simulation and the collision detection being created.

AR technology gave a lot of powerful features that assists un-expert workers during a robot programming tasks. Furthermore, it decreases the error risk during programming and increases the instructions' efficiency.

5.3 Path planning

Collaborative robot operations require a high level of accuracy and precision for paths and trajectory planning. One of the most difficult problems when programming a collaborative robot using AR technology is the teleoperation of the robot hence planning the robot's path [8]. The first proposed research in path planning and end-effector orientation using AR was in Singapore in 2012 [54]. The suggested approach, shown in Fig. 12, aims to create a path from a series of collision- free-volume (CFV). Then the path is optimized throughout a convex optimization method in order to get the best time- trajectory. Finally, a virtual robot and the planned trajectory are being simulated.

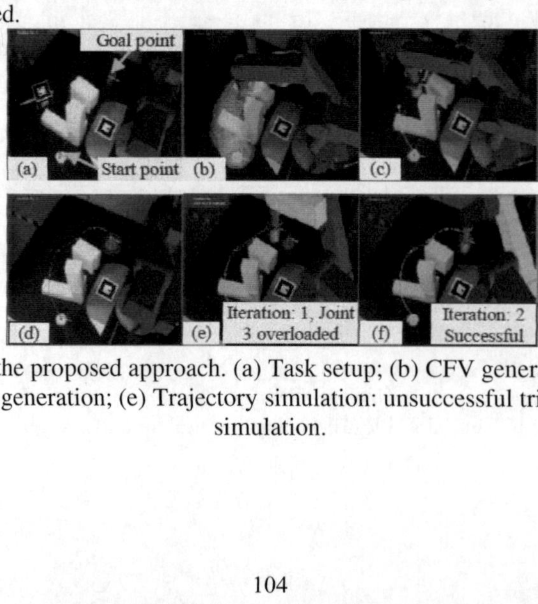

Figure 12. A case study of the proposed approach. (a) Task setup; (b) CFV generation; (c) Creation of control points; (d) Geometric path generation; (e) Trajectory simulation: unsuccessful trial; (f) Successful trajectory simulation.

Researchers tested this method for a pick and place task, as shown in Figure 12. (e) and (d), but this work is for virtual robots and not real ones.
Two years later, in 2014, the same research group presented a Euclidean distance-based method [55].

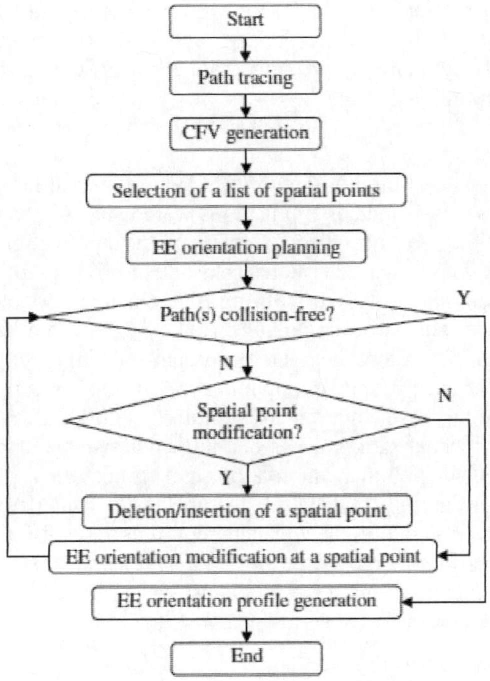

Figure 13. Procedures for the proposed AR-based path following operation [55].

Figure 13. illustrates the use of the CFV method in the suggested approach. First, a virtual sphere and a CFV area are created by recording the movement of the interaction device. Then, a parametric trajectory is being recorded by moving the interaction device. Later, the Euclidean distance-based method is used to the point of interest from the spatial points forming a collision-free trajectory. When the robot path is generated, the CFV verify if the EE of the robot still along the CFV during the movement. This approach gives also the possibility to get a collision-free trajectory for all the kinematic system of the robot. The experiment results of this approach for a pick-and-place task approve its benefits as a teach-in method and that it offers a bigger accuracy.
The next AR-based robot path planning was suggested in 2016 by C. Chu [56]. The proposed method aims to promote the robot path planning using AR for programming by a demonstration. First, the real objects are captured, analyzed and processed throughout an external vision device. Then, the results are combined with provided virtual objects. Last, the system launches a real-time visual information when detecting collision with an obstacle or an object.
In 2017, M. Tessarotto [57] presented an industrial robot path planning using AR. This method was especially focused on AR for offline more than online programming and, as all AR- based robot programming and path planning methods, it makes it feasible and easy even for inexpert operators. Record the trajectory is the first step of this method. An operator records the trajectory via the software interface or by an HTC Vives controller for an online robot programming. After recording the trajectory, it could be virtually displayed and modified throughout the program. The possible modifications are: "Distance filtering" that removes some points from the trajectory to smooth out the movements, "Flattening" to change every point of the path to smooth it itself, "Offset" to move a trajectory in order to be used in different context and "Only reachable" that enables the specification of whether the trajectory must contain all the points or to exclude the points not reachable by the robot. Later, a series of 'position/time' or 'position/speed' trajectory forms couples are required in order to create the robot program to be sent to the controller.
Recently in 2018, the "ARTool Zero" framework [58] was presented to support users during a trajectory planning. It also enables the generation as well as the simulation of the part- program that guides a touching

probe to identify the necessary geometric features. ARTool Zero displays the coordinates throughout a mobile device to the machine using markers that aims to convert the 2D plane into a 3D plane in the space.
Most of the existing robot path planning methods are time-consuming and may require sophisticated software tools which means expert engineers to develop the programs. New user- friendly AR-based path planning features are making this operation easier and possible for even inexpert workers. The majority of the presented works use AR for visualization of the suggested paths and collision-free trajectories. Quite a few, but increasing, number of AR-based path planning programs that allow efficiently and easily program trajectories all-alongside simple human intervention.

6. Conclusion

This literature review discusses the evolution of the use of AR in industrial factories since 2012. Based on the projects presented in this paper, several areas of applications were using AR technology. However, the type of utilization was evolving over time. At the outset, almost all the applications were for data visualization including simulation and design which means visualization of 3D products or special information related to real objects. Then it was used to support workers during maintenance operations or assembly/dis-assembly where the AR system displays assistant information generated by the system itself. In the meanwhile, remote assistance operations have appeared to remotely assist technicians on site by expert engineers located far from the factory. Remote assistance is used mainly for maintenance tasks and unexpected machine breakdown. Besides, along with the increasingly existence of collaborative robots, AR is being used for human-robot interaction. Similarly, it starts for visualization tasks, then human assistance. Until recently, AR went forward in industry 4.0 by being used in robot manipulation and trajectories planning. Although its recent implementation for robot programming, AR-based features still can be more powerful and requires development in this area. Moreover, all along the advancement of collaborative robots and their important role in factories, HRI and tasks operation should be feasible by any worker and as easy as possible. This will decrease the costs and the time of programming the robots as well as increase the efficiency and effectiveness as well. Thus, more research still needs to be done in this context.

Acknowledgment
This research is supported by the project Robotix Academy, within the programme INTERREG A – Greater Region

References

[1] M. A. K. Bahrin, M. F. Othman, N. H. N. Azli, and M. F. Talib, J. Techno (2016)
[2] B. Bunse, GTIA, (2013)
[3] G. Kipper, J. Rampolla, Augmented Reality: an emerging technologies guide to AR, (Elsevier, 2012)
[4] A. Balog and C. Pribeanu, Handbook of Research on 3-D Virtual Environments and Hypermedia for Ubiquitous Learning, 537-554, (IGI Global, 2018)
[5] P. O'Shea and C. Campbell, Virtual and Augmented Reality: Concepts, Methodologies, Tools, and Applications, 1644-1658 (2018)
[6] D. Andersen, V. Popescu, M. E. Cabrera, A. Shanghavi, B. Mullis, S. Marley, G. Gomez, and J. P. Wachs, *Military medicine,* 182, 310–315 (2017)
[7] K. J. Carlson and D. J. Gagnon, *Clinical Simulation in Nursing*, 12, 123–127 (2016)
[8] M. Billinghurst, A. Clark, and G. Lee, *Foundations and Trends in HumanComputer Interaction*, 8, 73–272 (2015)
[9] V. Krevelen, D. W. F., and R. Poelman, J. of VR 9.2 (2010)
[10] S. Mann and W. Barfield, J. of HCI, 15, 205– 208 (2003)
[11] J. Novak-Marcincin, J. Barna, M. Janak, and L. Novakova-Marcincinova, *Procedia Computer Science*, 25, 23–31 (2013)
[12] D. Mourtzis, E. Vlachou, N. Milas, and N. Xanthopoulos, *Procedia CIRP*, 41, 655– 660 (2016)
[13] D. Mourtzis, V. Zogopoulos, and E. Vlachou, *Procedia CIRP*,63, 46–51, (2017)
[14] H. C. Fang, S. K. Ong, and A. Y. Nee, *Advances in Manufacturing*, 2, 275–288 (2014)
[15] Z. Pan, J. Polden, N. Larkin, S. Van Duin, and J. Norrish, ROBOTIK (2012)
[16] A. Nee, S. Ong, G. Chryssolouris, and D. Mourtzis, *Manufacturing Technology*, 61, 657–679 (2012)
[17] V. Paelke, ETFA (2014)
[18] B. Lo¨fvendahl, Augmented Reality Applications for Industrial Robots, 167, (2014)
[19] I. L. Madar, M. Smith, and P. Knackfuss, LNICST, 150, 151–156 (2015)

[20] M. Fiala, TPAMI (2010)
[21] G. Caruso, M. Carulli, and M. Bordegoni, *Computer-Aided Design and Applications*, 12, 86–95 (2015)
[22] J.R. Chardonnet, G. Fromentin, and J. Outeiro, *Multimedia Tools and Applications* (2017)
[23] S. Su¨to, Z. Forgo´, and F. Tolvaly-Roca, *Procedia Engineering*, 181, 503–508 (2017)
[24] D. Mourtzis, V. Zogopoulos, I. Katagis, and P. Lagios, *Procedia CIRP* (2018)
[25] Y. Zhang and T.-H. Kwok, *Procedia Manufacturing*, 26, 1278–1286 (2018)
[26] R. Palmarini, J. Erkoyuncu, R. Roy, and H. Torabmostaedi, *Robotics and Computer-Integrated Manufacturing*, 49, 215–228 (2018)
[27] G. Dini and M. D. Mura, *Procedia CIRP*, 38, 14–23 (2015)
[28] R. Palmarini, J. A. Erkoyuncu, and R. Roy, *Procedia CIRP*, 59, 23–28 (2017)
[29] S. Webel, U. Bockholt, T. Engelke, N. Gavish, M. Olbrich, and C. Preusche, *Robotics and Autonomous Systems*, 61, 398–403 (2013)
[30] D. Mourtzis, V. Zogopoulos, and E. Vlachou, *Procedia CIRP*, 63, 46–51 (2017)
[31] C. Kollatsch, M. Schumann, P. Klimant, V. Wittstock, and M. Putz, *Procedia CIRP*, 23 (2014)
[32] A. W. Yew, S. K. Ong, and A. Y. Nee, *Procedia CIRP* (2017)
[33] D. Aschenbrenner, M. E. Latoschik, and K. Schilling, ACM VRST '16 (2016)
[34] C. Hub, D. Hub, K. Base, and F. O. R. Managers, *Service in industry HUB* (2017)
[35] X. Wang, S. K. Ong, and A. Y. Nee, *Advances in Manufacturing* (2016)
[36] M. M. Chang, S. K. Ong, and A. Y. Nee, *Procedia CIRP* (2017)
[37] M. Funk, T. Kosch, R. Kettner, O. Korn, and A. Schmidt, i-KNOW (2016)
[38] O. Danielsson, A. Syberfeldt, M. Holm, and L. Wang, *Procedia CIRP* (2018)
[39] M. Fiorentino, A. E. Uva, M. Gattullo, S. Debernardis, and G. Monno, *Computers in Industry* (2014)
[40] J. Zhu, S. K. Ong, and A. Y. Nee, J. Comp. Integ. Manuf. 28, 213–225 (2015)
[41] R. Masoni, F. Ferrise, M. Bordegoni, M. Gattullo, A. E. Uva, M. Fiorentino, E. Carrabba, and M. Di Donato, *Procedia Manufacturing*, 11, 1296–1302 (2017)
[42] J. Rambach, A. Pagani, M. Schneider, O. Artemenko, and D. Stricker, *Computers*, 7, 6 (2018)
[43] E. Ruffaldi, F. Brizzi, F. Tecchia, and S. Bacinelli, *Augmented Reality, Virtual Reality, and Computer Graphics* (2016)
[44] G. Michalos, P. Karagiannis, S. Makris, O¨. Tokc¸alar, and G. Chrys- solouris, *Procedia CIRP*, 41, 370–375 (2016)
[45] S. Makris, P. Karagiannis, S. Koukas, and A. S. Matthaiakis, *CIRP Annals - Manufacturing Technology*, 65 (2016).
[46] S. Galen, D. Liedtke, and D. Schilberg, *Human Interface and the Management of Information*, 10273, 161–169 (2017)
[47] O. Danielsson, A. Syberfeldt, R. Brewster, and L. Wang, *Procedia CIRP*, 63, 89–94 (2017)
[48] P. Leit, I. Mal, and D. Sedl, *Industrial Informatics (INDIN)*, 14, 176–181 (2016)
[49] P. Tsarouchi, A. Athanasatos, S. Makris, X. Chatzigeorgiou, and G. Chryssolouris, *Procedia CIRP*, 55, 1–5, (2016)
[50] S. Stadler, K. Kain, M. Giuliani, N. Mirnig, G. Stollnberger, and M. Tscheligi, RO-MAN (2016)
[51] H. Liu and L. Wang, *Procedia Manufacturing*, 11, 22–30 (2017)
[52] S. Maksymova, R. Matarneh, V. V. Lyashenko, and N. V. Belova, J. of Comp. and Comm., 05, 1–15, (2017)
[53] J. Wassermann, A. Vick, and J. Kru¨ger, *Procedia CIRP*, vol. 76, pp. 161–166, (2018)
[54] H. C. Fang, S. K. Ong, and A. Y. Nee, *Procedia CIRP*, 3, 191–196 (2012)
[55] H. C. Fang, S. K. Ong, and A. Y. Nee, J. Inter. Des. and Manuf., 8, 33–42 (2014)
[56] P.-c. Li and C.-h. Chu, A-DEWS, 1–5 (2016)
[57] M. Tessarotto, Trajectories Programming for Industrial Robots Through Augmented Reality (2017)
[58] M. Ragni, M. Perini, A. Setti, and P. Bosetti, *Computers and Industrial Engineering* (2018)

Institute of Physics
Dirac House, Temple Back
Bristol BS1 6BE UK

ISSN: 1757-8981
ISBN 978-1-5108-8739-8
9 781510 887398

CURRAN ASSOCIATES INC.
proceedings.com